THE LIVE EVENT TECHNICIAN

The Live Event Video Technician covers terms, format types, concepts, and technologies used in video production for corporate meetings, concerts, special events, and theatrical productions.

The book begins by providing a history of the industry and an overview of important roles and functions therein. It then discusses various display technologies such as LED walls and video projection, as well as video systems for converting and switching of various types of sources. Presenting the cornerstone formats, connectors, and methodologies of visual technology, this book offers a strong foundation to help readers navigate this ever-changing field. Written in an accessible tone, the book clarifies jargon and is an overarching source of knowledge for the role of the video technician, for which there has previously been little formal training.

The Live Event Video Technician provides a wealth of practical information for students of media and communications courses, readers with a novice or entry-level understanding of video and AV production, and anyone with an interest in working as technical personnel in live event production.

Tim Kuschel has been a video engineer in the AV and Live Event industries for over 30 years, having done shows and events for clients such as the NFL, MLB, the Walt Disney Company, Microsoft, and Cisco Systems. Holding a certified technology specialist designation, he has developed certification courses products and taught hundreds of technicians worldwide. He is currently the director of visual technologies for OSA in Las Vegas, and a member of the Society of Motion Picture and Television Engineers (SMPTE) and the Audiovisual and Integrated Experience Association (AVIXA).

THE LIVE EVENT VIDEO TECHNICIAN

Tim Kuschel

Focal Press
Taylor & Francis Group

NEW YORK AND LONDON

Cover image: IATSE Video Technicians working at the Barco
EC-210 console. Photo by Tim Kuschel.

First published 2023
by Routledge
605 Third Avenue, New York, NY 10158

and by Routledge
4 Park Square, Milton Park, Abingdon, Oxon, OX14 4RN

Routledge is an imprint of the Taylor & Francis Group, an informa business

© 2023 Tim Kuschel

Library of Congress Cataloging-in-Publication Data
A catalog record for this book has been requested

ISBN: 978-1-032-16097-9 (hbk)
ISBN: 978-1-032-16095-5 (pbk)
ISBN: 978-1-003-24703-6 (ebk)

DOI: 10.4324/9781003247036

Typeset in Joanna
by Apex CoVantage, LLC

To my wife,
Maria-Victoria
and our kids,
Brandon, Kristen, Cameron, and Natalie,

thank you for all the love, support, and understanding all these years that allowed me to grow in my career and be able to have adventures with you.

To Bob Murdock,

thanks for being the other half of me all these past decades. My cheerleader, cohort, and my friend. For believing in me when I sometimes didn't.

CONTENTS

PREFACE

As of this writing, the world has been engulfed by the COVID-19 pandemic of 2020. In March of this year, as the number of reported cases and deaths were increasing, the live events industry began to see work recede at an alarming pace. Within weeks, major industry conventions, concerts, and corporate events of all sizes were beginning to cancel their bookings. This meant that all the associated production companies and technical personnel were also receiving cancellations. In some cases, this happened as they were loading in events or were already fully set up and ready for rehearsals.

Within three to four months, 99.9% of all live events for the rest of 2020 were no longer happening. This went on to include concert tours and stage productions—basically, anything that had an audience of more than ten people. Hundreds of entertainment and technical professionals found themselves unemployed with no timeline for when their industry might be allowed to resume normal operations.

The popular buzzword became "pivot." How can companies pivot their technology into virtual events? How can I pivot my skills into another industry? The unfortunate answer is that not many companies or people were able to pivot, at least not to a level where there would not be sustained losses. There was a trickle-down effect. Rental staging companies were not moving equipment and not in any position to buy new gear. This meant not having work for their in-house and freelance staff. Manufacturers of products in the live event market experienced a lack of sales, causing reduced staffing and cutbacks to production. Likewise, their supply chain was affected and so forth.

Those larger staging companies with the space and the equipment began building virtual studios. They were mainly greenscreen spaces or LED walls to which any manner of location content could be intermixed with live talent. If your people couldn't gather in the real world, let them tune into you on a live internet stream. Technology wise, nothing too new here. We had seen these types of virtual sets being used for news, film production, and sports programming.

So, what does all this mean for readers of this book? The answer is that hopefully this is now in the past, and the live events industry is on the mend. In terms of technology, HD (high definition) will be our starting point. There is enough literature on the history of television to not have to repeat the events leading up to the digital transition of the late 1990s and early 2000s. What there isn't too much literature on is the evolution of the AV industry and how, in particular, video has been a hybrid discipline, having one foot in the broadcast world and the other in the realm of high resolution displays.

If this book is your embarkation point into a new career, I wish you all the best and I hope you find it informative and entertaining.

1

LET THERE BE LIGHT . . . CANNONS

The history of video projection and displays for live event production spans more than nine decades—from the first broadcast of the 1938 Olympic Games in Berlin, Germany, to the use of video projection and video cube walls in the 1980s, to modern day LED extravaganzas driven with content from media servers.

For our starting point, we will use the mid-1980s, when AV presentations on a large scale were performed with stacks of 35 mm slide projectors. Presentations were coordinated to a timecoded soundtrack with cues programmed and triggered from early computers running DOS based control software. Slide jams or adding or removing slides from the presentation was a major undertaking.

What about moving images? The illusion of movement was created with animated slide sequences where projectors were triggered in rapid succession to create the flight of a bird or other objects in motion. There may also have been blank slides programmed into the show where a

DOI: 10.4324/9781003247036-1

Figure 1.1 AV engineers at Ford Automotive are programming a multi-image slide show using AVL Eagle computers in 1987.

16 mm film projector might be used to fill the area with motion imagery. These shows took days to setup and align and months of planning, storyboarding, image production, and programming to execute. There were little to no last-minute changes, as that would have meant having to find an overnight film processing lab and equipment for computer-to-film transfer. But what does this have to do with video?

As these shows continued, the film projectors were replaced by CRT (cathode ray tube) three-color video projectors. Now video clips, which were faster and easier to create than film, could be incorporated into multi-image presentations. Later, computer-to-video interfaces would allow computer imagery to be projected in these shows. Although these wide format type of presentations would fade out in favor of multiple projected single-video screens, the concept of ultrawide blended images would rear its head again in the future.

The CRT projectors were reliant on a phosphor-based image, but the technology had brightness limits. The lights in a meeting room generally had to be completely out to view images clearly. By today's lumen measurements, these projectors were in the range of 400–800 lumens.

But a technology existed for decades before that was used mainly for military and scientific applications. This was light valve technology.

The concept here is that an electron beam strikes an oil covered disk. The image created on the disk combines with the light source (originally arc light and later Xenon) and travels out the lens to the screen. An oversimplification, but the goal of this technology was to produce video images that were bright enough for screens 30–40 feet in width. The first projector to use this method was the Eidaphor (nickname IDA4). Developed prior to World War II by Dr. Fritz Fischer, the Eidaphor would go on to produce the large video screen images at NASA's Houston control center that tracked the first manned flights to the moon. It was also, probably the projector behind displays in wartime situation rooms and NORAD, although that hasn't been confirmed or denied.

Through the late 1970s and into the 1980s, music artists began to expand the concert going experience to include more visuals, including video. David Bowie used projected video for both his "Glass Spider" and "Sound+Vision" tours, in 1985 and 1987, respectively. The only technology bright enough would have been light valve. But the Eidaphor had issues in the live event world. The refrigerator-sized machine needed to be steady and was built heavy to reduce incidental motion. Any disturbance of the oil would cause an "image amoeba" that would take about ten minutes to dissipate. Its components needed to operate in a vacuum and the oil had to be at a certain temperature. This meant once the unit was setup and operating, it should not be turned off.

However, if you wanted big and bright video images, you could only rent an Eidaphor from military contractor SAIC and they supplied a technician.

Then along came the GE Talaria light valve projectors. These were based on similar imaging technology but were built more specifically for event use and traveling. They still required a good deal of setup time, but now technicians in the private sector were able to get training on their operation and AV rental companies could own the projectors.

Another way to create large video images during this time was with video cube or projection cube walls. A cube wall was generally a stack of CRT television monitors in some matrix of 3 × 3, 4 × 4, and so worth. These were special monitors with the ability to adjust the convergence, either manually or electronically, so that images could be aligned across horizontal rows and vertical columns. Since CRTs used magnets for beam control, every time they were set up a technician had to use a degaussing ring to reduce or eliminate the magnetic field caused by the neighboring monitors. These would manifest as hot spots of colors on the display.

Projection cubes were CRT projectors mounted at one end of a four- to five-foot long square tube. At the other end was a rear projection surface. The projector (referred to as the engine) could be removed after the matrix wall was assembled for replacement or maintenance. Projection cube walls were preferred over monitor walls since their images could be edge-butted together to create a "seamless" overall image. Monitor walls had gaps, called bezels, that were spaces in between displays that broke up the imagery with a visual grid.

A staple of the early MTV years, video walls were used everywhere in the early 1990s, from shopping malls to major concert tours like U2's 1992 "Zoo TV" tour, which took video for live events to the next level combining projection, video walls, and single monitors into a sensory overload experience.

Control and imaging for videowalls was handled by a video wall processor. Various video sources could be input, and software control allowed the images to appear individually on every monitor or spread across the entire display. This was programable as presets so different looks could be recalled. This made video walls an interactive video element that could be more than just a large video display. Their major downside was they were time and labor intensive.

In the mid-1990s, the Belgian-based company Barco introduced the first of their light cannon projectors, the Barco LC 5100. Barco had been around for some time in the projection market, having created some the brightest and highest resolution CRT projectors. One thing that made the LC 5100 different was its imaging system. It used three LCD panels that were optically converged out a single lens and its light source was a single

UHP (ultra-high performance) mercury-halide lamp. The output rating was 1500–2000 lumens, which was bright considering CRT projectors were not capable of even 1000 lumens. The lens was interchangeable to allow for different zoom ratios, however the lens had to be removed when shipping the projector by screwing and unscrewing it from the lens mount.

What the light cannon brought was the beginning of what could be called "point and shoot" projection. Although still bigger in footprint than most projectors today, it was less cumbersome and required less setup time than the GE Talaria light valves and a double stack of light cannons could produce just as bright an image. Operators of the Barco Light Cannons still had to adjust panel convergence at each location because the alignment system was based on a spring tension system. Jolts or rough handling in shipping could misalign the panels.

Barco went on to manufacturer brighter light cannons with their 9300 series boasting a whopping 5000 lumens. What also was being developed in parallel was smaller LCD based projectors by companies like Sharp, Eiki, InFocus, and Panasonic, that were smaller and could outperform legacy CRT technology. Generally, there was no regular convergence requirements for these newer LCD units since they used only one full color panel. As with any technology, the size and weight of these LCD projectors reduced as brighter versions were released.

Then in the late 1990's, a projection system based on the Texas Instruments DMD (digital micromirror device) was released. It was called DLP, or digital light processing. One of the first units was made by Digital Projection and was rated at 5000 lumens. What made DLP groundbreaking was that the "light engine" was one assembly. There was little need for DMD panel convergence and the color representation was on spec with film and broadcast color calibration standards. You could literally turn the unit on, point, focus, and shoot.

The downsides were a, huge form factor and a xenon lamp light source that required a minimum of 208 volts of power to operate. When the lamp blew, it was as loud as a 12-gauge shotgun. You then had to wait for the unit to cool down before you could remove the blown lamp and vacuum all the glass dust and debris before inserting the new lamp.

But DLP would progress with the release of the Electrohome Roadie. A 10,000-lumen projector that was built for road travel and included hardware for double stacking units and had a side removable lamp module. Electrohome would soon be acquired by Christie Digital and would expand the line with the S12 (12,000-lumen) Roadie and smaller profile units called the Roadster series. Christie was the largest manufacturer of film projectors for cinema theaters. Their acquisition of Electrohome was part of a strategic plan to enter the rental staging video market and develop DLP for digital cinema adoption.

Other than minor competition from Digital Projection, Christie Digital ran unopposed in the DLP large format video sector for several years. It wasn't until Barco realized they were losing projection market share by depending on LCD that they retooled and began producing DLP based units in the early 2000s. The dominance of Barco and Christie Digital in this market would last for more than a decade.

Since the DMD technology was owned by Texas Instruments, they controlled the resolution of the chips made available and set limits on how bright the light could be that was exposed to the imagers. Projector manufacturers were responsible for housing design, optical path, and image processing of signals. As such, images from different DLP projectors became somewhat indistinguishable with only brand loyalties determining selection. What did sway decisions were the manufacturer designed elements such as on-board menus, remote network operation, stacking and rigging hardware, lens options, and feedback from technicians in the field.

More than 20 years later, DLP is still the primary technology used in large format projection. Other technology, such as LCD, still exists and is in use. The details of these systems and light sources will be covered in a future chapter.

2

PIXEL WRANGLING

Live event video has always borrowed technology from the broadcast world. From cameras to switchers to routing systems, reutilizing the tools for television production made sense since projectors and displays are often designed around current broadcast formats. Looking again at the Eidaphor light-valve projector, all its weight and complexity was designed for the 525-line NTSC (National Television Standards Committee) signal or 625-line PAL (phase alternating line) formats, which were the standards for most of the world from the 1940s until the new millennium.

With the introduction of computers in the home and workplace in the 1980s, it would be inevitable that a database, spreadsheet, or CAD design would need to be shared on a large screen. This is where computer interfaces created by names like Extron, Inline, and Covid would convert the DB-15 VGA analog output of a computer into five wires carrying red, green, blue, horizontal sync, and vertical sync, commonly referred to as RGBHV. This format conversion was necessary as the analog VGA output of

DOI: 10.4324/9781003247036-2

a computer was designed to go a short distance, from the computer to the monitor sitting on the desk. RGBHV allowed for the signal to be amplified and distributed to multiple displays over long distances. However, a 100-foot Canare V5 5-wire cable was about an inch and a half in diameter and weighed about 40 pounds. Forget pushing over tractor tires in the gym, try carrying those cables up multiple flights of catwalk stairs.

The resolutions of these early PCs started at 800 × 600 pixels and eventually became the more popular 1024 × 768. Most "data" projectors and monitors could handle these higher rates and the match of a 4:3 aspect ratio—the same as video—meant they could be multipurposed. This made them more expensive and slightly heavier than their video only counterparts.

But what if you wanted to have a presentation involving computer and video sources? At first, your only option was to switch between inputs at the projector. You would see the projector go to black and then return with the new source. For CRT projectors, they had memory systems based on the pixel clock of the resolution. This meant you had to converge the projector for the expected signals during your presentation. It was very common to require presenters to connect their computers the night before or prior to the presentation, so that the technician could converge the projector's image to the signal. If the presenter didn't show up, they risked having their image display misaligned. Some manufacturers had extended memory files in their projectors that could be loaded in between presentations or external accessories that would switch from one signal source to another without the dip to black.

What was needed was a way to have all signals be the same resolution so that the projectors only had to process one resolution. One solution was to line double standard video signals to be closer to computer resolutions. Faroudja line doublers were the premiere processors used for this purpose. By doubling a 525-line NTSC video signal, you could come close to the pixel clock rate of computers. This meant a shorter switching time between sources as they could use the same memory file.

The Faroudja line doublers were very popular with the high-end home theater crowd. In fact, they were among the few that could afford the price of the processor. Additionally, you had to have high-quality video source material to appreciate the processing. After all, crap in equals two times crap out.

Going the other direction, there were manufacturers producing computer graphic cards that could output in standard video formats and could be genlocked (genlock will be discussed later, but it is basically a timing signal that video devices use to synchronize their outputs). Two of these hardware developers were Targa and Matrox.

These graphics cards got their start in character generators (CG) for television production. CG units produced the titles, lower thirds, credits, and weather graphics that were overlayed or keyed onto camera images. In some cases, they could display the computer's desktop. But these images were generally unreadable as the computer natively was running at a higher resolution than the output of the video graphics card. They were also impractical for most corporate events due to the fact that the presentations had to be created entirely in advance and they required special programs that required significant skill to create content.

Enter the video scaler. A chip-based processor that could remap one resolution into another. One of the pioneers of this technology was Folsom Research. Located in Folsom, California, they began as a manufacturer of radar displays for air traffic controllers. One day they were tasked with taking the circular sweep of the display and "rasterizing" it to be shown on a standard monitor. This research led to developing their proprietary Athena scaler, named after the owner's daughter.

One of the groundbreaking products to use this scaler was the Folsom VFC-2200. VFC stood for video format converter. It had two scaling channels and the user would set the desired fixed output resolution (e.g., 1024 × 768). Incoming standard video would be scaled up to the 1024 × 768 computer resolution. The display device, such as a projector, only sees a single constant format signal. All switching and scaling of signals happens upstream and source-to-source transitions dissolve seamlessly (heavens open and angels sing). One of the flaws in this early technology were oscillators that were fixed at 50 or 60 Hz. Video in the United States actually cycles at 59.94 Hz. So, about every 12 minutes or so, as the 60 Hz and 59.94 Hz signals drift far enough apart, the image would "jutter" as they realigned. 'The Jutter Effect' as it is called, causes live video to freeze or stutter briefly.

Since the VFC-2200 only had one input per scaler, systems were devised involving signal routers upstream of the scalers to handle large numbers of input sources. At the same time, memory files for each of the different sources needed to be recalled on the VFC-2220 when the input was changed. Enter the screen master controller developed by Clark Williams of Vista Systems. Folsom Research knew how to make scalers, but control surfaces were not an area of expertise. Through self-study, Clark developed a control surface that managed router commands for signal switching and recalled saved memory files for sources stored in the VFC-2200. Most importantly, the ability to save all these parameters as a preset, a recallable cuing system for single or multiple screen destinations.

As the number of screens needing to be controlled increased, so did the control surfaces. Before projectors had adjustments for edge blending and image feathering, those controls were in the VFC-2200 and the Screen Master controllers allowed users to position PIP windows in real time and save their locations in presets. Users now had the ability to preview their screen looks before committing them to screens, just like they had been able to do on broadcast production switchers up to this point.

Folsom Research and Vista Systems partnered for several years until Clark Williams decided he was going to build his own scaler and screen control system called Montage. This would later evolve into the X20 Spyder system which would directly compete with the next generation screen management system from Folsom Research called Encore. These two high-resolution switching systems were at the heart of any multi-screen, widescreen blended projection, or fixed installations during the first half of the 2000s.

Eventually, Folsom Research was acquired by Barco, and Vista Systems was acquired by Christie Digital, the two main projection competitors in the rental and staging industry at that time. Barco would go on to develop the E2 Event Master and Christie the X80 Spyder.

Meanwhile, in the background of all this was a company based in France named Analog Way. Their product line would include units that had similar function to Barco Folsom's smaller switching line. In fact, Analog Way would have a much more diverse catalog and its products were

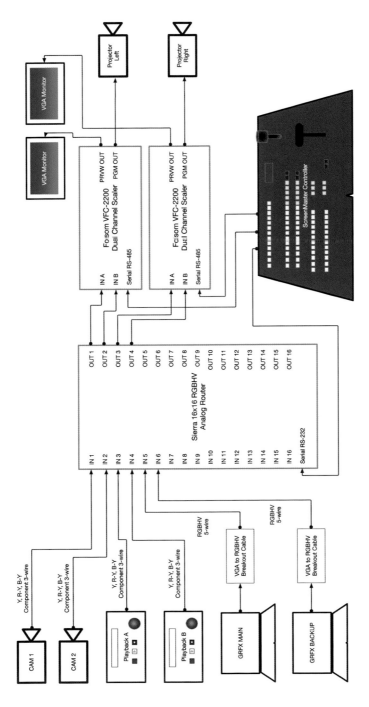

Figure 2.1 Signal flow diagram of a VFC-2200 Screen Master system.

Figure 2.2 Screen Master 3216 controller with Folsom VFC-2200 scalers controlling the image to blended widescreen projectors.

popular with the integration and house of worship markets. Eventually, they would develop the LiveCore processor which would be their basis for the Ascender series. These systems would compete with the Barco Encore and X20 Spyder systems. Analog Way evolved their technology as well and introduced the LivePremier systems to compete with the newer Barco E2 and X80 Spyder. Technology in this area has always been a leapfrog, with one manufacturer taking the spotlight for a few months to a year until a competing company could develop a product with more processing power and more features.

Mixing and Matching Formats

Besides being able to transition from one resolution to another, scalers allowed us to manipulate sources to conform to different standards, aspect ratios, and scan rates. With regards to aspect ratio, it was not uncommon to have to take a letter box video and force it to fill the full output raster of a standard definition output—the logic being, "Why are we wasting the space at the top and bottom of the screen?" Once you did it, you then got asked why the left and right portions of the image were missing. This led to having to explain that although all squares are rectangles, not all rectangles are squares.

Another feature that scaling technology brought was the ability to do format conversion between standards. Converting a PAL signal to NTSC or vice versa, or from a 1024 × 728 computer signal to a standard definition television signal, was now effortless. But what about from one medium to another, such as film to video?

The major issue between film and video is the translation of a frame rate into a scan rate or refresh rate. For decades film makers have been using the frame rate of 24 frames per second. However, video started out with a scan rate of 29.97 frames per second prior to the deployment of high definition. To transfer a movie shot on film to video in order to be broadcast, a machine called a telecine was used. The telecine recorded the film in real time as it played. The camera in the telecine would capture the frames and convert them to a video format using a special pulldown technique. This electronic manipulation, called 3:2 pulldown in the United States, would create an extra field for every third frame of film captured. Early television was interlaced, odd lines (fields) rendered followed by even lines (fields). This extra field allowed the film to playback properly at the broadcast 29.97 frame rate.

Fast forward several decades, and we now have a world where most display devices run at either a 50 Hz or 60 Hz refresh rate. Most films are still shot at 24 frames per second, even though we generally no longer use film as an actual medium. So why haven't filmmakers embraced 50/60 frame rates to make transfer to video easier? The reason is purely artistic. Film has a "look" that is destroyed by the hyper-realistic look of capturing at 60 frames per second. This may change as how we view movies changes. The only place to truly see a film in native 24 frame is in a movie theater; otherwise you are watching it on a display at 60 frames per second with pulldown conversion.

How does all this effect you as a live event video technician? You may have to work a film festival where there is a mix of 24-frame and 60-frame content. Does the equipment you're using handle 24 frame? Since many film shoots still happen in 24 frame, if you are working a concert event or other scenario involving LED video displays that are being filmed, can those devices function at a 24 Hz output so that there are no visual artifacts or banding?

3

THE PEOPLE IN YOUR PRODUCTION NEIGHBORHOOD

Like any company, there is a structure to the personnel on a typical show site. There is a chain of command and reporting structure that should be followed with regard to operation and conveyance of issues. Circumventing this may cause you to not be asked back or denied future opportunities.

End Client

This is the company that the event is being put on for. Their name is plastered all over the materials and signage. For corporate events, this person is typically the director or vice president of marketing or communications. Basically, someone within the company designated to oversee the event. In the case of concerts, this is the artist performing on stage.

DOI: 10.4324/9781003247036-3

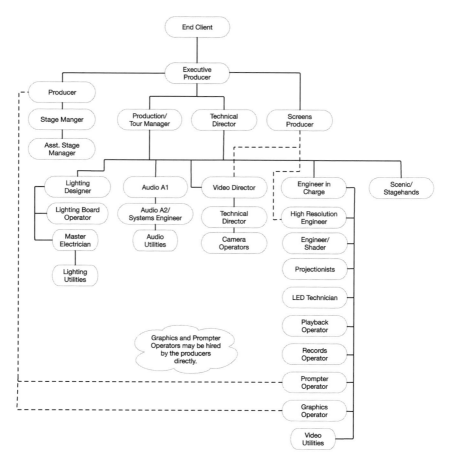

Figure 3.1 The hierarchical structure of typical live event crew positions.

Executive Producer

The end client has hired a production company to organize their meeting, book guest speakers, and take care of the technical details. The person from the production company that has direct access to the end client is the executive producer. This person has the final say with regard to budgetary matters and creative decisions. Those who interact with this person are limited to the cadre of producers brought in to handle various aspects.

Producer

The producer is in charge of the content for a given session or room. Their job is to ensure the show is executed according to plan and to coordinate the speakers and elements. They interact with the executives or talent in their venue and approve any requests or changes.

Screens Producer

On shows that involve blended widescreen projection, large LED walls, or a combination of both, a screen producer may be involved. Their responsibility is to design and organize the content as it will be presented to the audience. They will often work with the creative team to make sure the content is pixel accurate and see to it that elements are lined up in the correct locations. They will often produce storyboards showing the various scenes throughout the program.

Production Manager

This position could have two descriptions based on the event. For concert tours, the production manager (PM) is like the executive producer in that they are the highest level of authority short of the end client. Production managers are responsible for interfacing with venue management and handling the payment of local and touring crews as well as operational oversight.

For corporate events, the production manager may be a person assigned by the main equipment vendor to coordinate the installation and show personnel. When one company provides all disciplines (audio, video, and lights), having a PM for the upfront work is common.

Technical Director

The executive producer or their production company will either have a staff technical director (TD) or may hire one they have worked with over and over. The technical director is responsible for all the technical elements of an event. This means putting out to bid and hiring the audio, video, lighting, and scenic vendors. Generation of CAD drawings and site surveys will also fall under their area of responsibility. Additionally, they will be responsible for hiring local labor and coordinating crew meals with hotel or venue staff.

All equipment changes or additions need to go through the TD. They are the coordinator of load-in and load-out schedules between vendors and responsible for show site safety.

Stage Manager

Once the show or event is loaded in technical rehearsals begin, the stage manager takes over the operation of the crew and show flow. They are typically located at front of house (FOH), a platform behind the audience with a full view of the stage. Seated next to them is usually the producer. They run the cues and are the only ones allowed to say "go" during run of show.

Stage managers are provided with a multi-viewer display from the video department. This contains windows of the presentation support graphics, playback sources, and prompter feeds.

Assistant Stage Managers

These people are positioned in the wings of the stage or backstage. Their job is to wrangle executives and talent from green rooms and back areas to the stage. They direct stagehands with regard to scene changes and prop movement.

Audio A1

Also at FOH is the main audio operator in charge of mixing the show and all supporting audio elements. They will also control the levels for record feeds going to video. Records operators will have the most interaction with them for this reason.

A2/System Engineer

This person makes sure the speakers are placed and programmed correctly and miscellaneous audio feeds are taken care of. This includes the placement of audio monitors throughout the backstage and client areas. If not delegated to others, this person may also be in charge of intercom and wireless microphone management including frequency coordination.

Lighting Designer

The person responsible for the design of the lighting cues during an event. They specify the type and number of fixtures and generate a lighting plot in CAD.

Lighting Board Operator

Depending on the size of the event, the lighting designer may also be the lighting board operator. This person, another member on the FOH platform, physically programs the cues into the lighting console and operates the run of show. Video directors will interact with them to set light levels, and a video monitor with the ability to switch between camera feeds or the main program feed is placed at their position. This allows them to adjust how the lighting looks on camera. How lighting looks to the naked eye versus a camera can be very different. It is important for the person controlling cameras to communicate with the lighting department in terms of getting time to color balance cameras to the lighting.

Master Electrician

This position is responsible for making sure all the electrical and control wiring for the lighting department is carried out. Their area traditionally has been nicknamed "dimmer beach." With most intelligent lighting these days not requiring dimmers because they are LED based, this area is mostly made up of power distribution racks. Lighting generally makes up the brunt of the labor on load-in and load-outs due the amount of cabling, fixtures, and truss required.

Video Village

The video department will more than likely make up the bulk of operators during the run of show. They will take up the largest footprint of backstage area and hence their space has been nicknamed "video village." Not only does it take a village to raise a child, but also to run major shows.

Engineer-in-Charge

In live events, the engineer-in-charge (EIC) is responsible for the operation of all video systems on a show site. They oversee the patching and connections for signals related to cameras, recorders, stage displays, projectors, and wherever else a video feed may be required. In a broadcast scenario, they have technical responsibility for all departments.

This person is designated by the video vendor as they will have familiarity with the vendor's systems and have the authority to make changes or request additional equipment if needed. Communication between the EIC and the technical director prior to and during the event is critical.

Video Director

This person may or may not be assigned by the video vendor. In most cases, they are chosen by the production company based on work history. Given the size and complexity of the event, they may just call cameras and someone else will operate the video switcher, or they may to do both. Camera position determinations fall under this person, and they will produce what is called the "line cut,", cuts between cameras and graphic elements without dissolves.

Engineer/Shader

Helping the EIC with patching the system, the engineer/shader also makes sure that all the cameras are connected for image, tally, and intercom, and monitors the exposures of the cameras during run of the event. Color balancing of the cameras so that they all having matching images, is also important. This position communicates with the lighting department to ensure proper levels.

Technical Director

This position is very different from the show's technical director described earlier. To avoid confusion, this position is usually referred to as TD switcher or just TD. In Europe and other places, they are called the vision mixer.

This is the person sitting behind the video production switcher mixing the cameras and other sources. They produce what is called the "program feed." In broadcast events run from a production truck, this person is the final stop before the feed goes to air. They key in graphics such as titles and scores, and route feeds to auxiliary destinations and the multi-viewers.

In a concert or corporate event, their responsibility may be limited to just switching the cameras. The control of the screens and other display devices falls on the next position, the high-resolution engineer.

High-Resolution Engineer

With computers as a part of every live event, the high-resolution engineer is responsible for two aspects, signal cabling for high-resolution sources and feeds to displays, such as projection and LED walls. Rolled up into this position would be operating the screen management system, such as a Barco E2 or a Christie Spyder X80.

Division of tasks is necessary on any event. Expecting the TD switcher to give camera direction and cut the camera feeds, then simultaneously trying to listen to the stage manager give cues for screen looks, is too taxing on one person. By having the high-resolution engineer focus on what is on the displays and the cues the stage manager is giving, the director is free to produce a better line cut of the show.

Playback Operator

The position responsible for playback of any video media. In today's world, everything is in digital form. It is more common for someone to hand the playback operator a USB thumb drive rather than media recorded on Blu-ray disk or tape. This means the playback operator has to be familiar with common video file types, including .mpeg, QuickTime, and HAP. They will be playing this media using a computer-based system using any number of software programs. Knowing what format the software and hardware like best, and having a toolset of conversion programs, is key to this position.

Included under this position are media server operators. For shows with large pixel spaces or multiple-screen synchronized content, a specialized computer called a media server will be used. What differentiates a media server from regular playback is that playback is generally one single stream being played out, while media servers often have multiple output streams. Additionally, they include the ability to warp content or output to different size pixel spaces.

Records Operator

This position has one job: to press the record button. Records operators will also monitor the audio levels on the recordings and do file transfers from record media to larger hard drives for handoff to the client. They will also keep a log and note the time code of any anomalies in the recording such as audio dropping out or video quality issues.

In the current world where every show is streaming live to the internet, the start and stop of streaming feeds may also fall under their area of responsibility.

Graphics Operator

Proficient in Microsoft PowerPoint or Apple Keynote, the graphics operator is responsible for following the presenter's cues and forwarding their slideshow as necessary. Other tasks may include simple editing or reordering of slides. Heavy editing of a presenter's presentation is usually not the responsibility of the show's graphics operator. Typically, on site there will be a graphics editor who will handle these situations in a speaker rehearsal room. However, having above average skills with PowerPoint will always be a plus and something that may place you in higher demand.

Prompter Operator

Guiding the presenters through their speech is the prompter operator. You may have seen images of scrolling text in front of cameras on news shows. The anchorperson is reading pre-written text while looking at the camera versus reading off papers on the desk.

In live events, the concept is very similar. Monitors down in front of the presenter, called DSMs, or down stage monitors, present the rolling text for the presenter on stage to read. Prompter operators have to become adept at following the cadence of a person's speech and either leading them or slowing them down by controlling the speed of their scrolling.

Projectionists

This position is responsible for the setup, alignment, and daily operation of any projection systems on the event. Not only do they need to be familiar with the controls and features of the projectors that they are using that day, but also cognizant of the projectors weight and safety considerations when rigging the projectors above the audience.

Having a fear of heights would not bode well in this position as projectionists are often in scissor lifts or climbing scaffolding to make adjustments.

LED Technician

LED technicians are responsible for the LED displays on the event. They lead the construction, wiring, and programming of the mapping of LED tiles.

Camera Operator

These people operate the various types of cameras that may be on a live event. Static, or hard cameras, are any camera position that will not change location. They usually are at the back of the room and may use long range zoom lenses. Handheld cameras are mobile and are usually operated closer to the stage. Other camera positions could include jibs or crane cameras which swing large arms out over the audience. This takes skill and practice, and most jib operators own their own equipment due to its specialization.

Generally, employment in the live events industry falls into one of three categories: staffer, freelancer, or union member. A staffer is a person who works directly for the production company and gets paid by them as a full-time employee. The benefits to being on-staff are a regular paycheck, guaranteed work, health insurance, and other benefits, along with being able to work with the same team and familiar equipment each time.

Freelancers are contract employees that are hired per event by either the show's technical director, a production company, or an equipment vendor. They must carry their own health and liability insurance, and sometimes workers' compensation coverage. Freelancers can work for several different companies and have the ability to pick and choose which events to work or take no work in order to have breaks or vacations. Most freelancers form their own independent companies to manage liability and for tax advantages. Others book through labor brokers that will process their payroll for employment taxes and will provide liability and workers' comp coverage.

Union members will be employees of a local IATSE (International Alliance of Theatrical Stage Employees). Once signed up, you will be dispatched on events based on your skills or experience. The structure of the union allows for growth through apprenticeships and journeyman programs, and seniority determines priority on bookings.

Being a member of the IATSE can provide access to group health benefits, retirement plans, and job protections through the grievance process and collective bargaining.

It is not uncommon for all these types of employment situations to be working on the same show site. Production companies use a combination of their staff employees along with freelancers to operate their equipment, along with local stagehands to get the gear into the room and help with set up.

There are also situations where the union can have jurisdiction over a property and all operators must be union members. In those cases, any operators that are not union members must have a "shadow," or person that is billed by the union for the exact same position to the client. Generally, the person shadowing an operator should be capable of taking over that position if necessary.

Life on the Road

If you get the unfortunate reputation of being good at what you do, you'll find yourself often traveling to destinations around the world to do a job that someone locally is perfectly capable of. Why is that? Because a client or production company has become comfortable with you and can rely on you for consistent and successful show every time.

A good freelancer usually has two or three clients that keep them busy all throughout the year. Ideally you want to work about 200 days a year on shows. Know what your position should be paid and be wary of taking positions you know are paying less than that.

Be sure to sign up for the rewards and mileage programs associated with airlines and hotel chains. With most airline programs, the person actually flying earns miles toward rewards regardless of who paid for the ticket. With hotels it's a little different. Credit for stays only happens if you are actually paying room and tax. If it's being billed to a third party, such as the show's end client, then you will only get reward points for incidental charges incurred during your stay, such as hotel meals and parking.

If you are going to be a frequent traveler, either be self-reliant or have a strong support system that understands what you do for a living. Relationships can be trying in our industry, especially when it comes to raising families. Today, technology helps with the ability to Zoom call or FaceTime with the people at home. But nothing replaces one-to-one time, and don't forget that scheduling family time is just as important as scheduling shows.

4

IN THE MIDDLE . . .

In the middle, there was HD. As part of the "analog sunset" of the late 1990s, broadcasters were transitioning to 16:9 aspect ratios and digital over-the-air transmission. Two formats were floating around, 720p and 1080i, and it was left to production companies and television stations to decide which way they wanted to go.

Some, like Fox Networks, chose 720p as their HD flavor of choice. It allowed them to have a progressive output which they felt more aligned with film, and the 1280×720-pixel resolution wouldn't require as much storage space as 1920×1080. Remember, in the late 1990s memory was not as cheap as it is now. Likewise, it also required less bandwidth for file transfer and distribution.

But despite whatever format a television studio or production company produced the material in, it went out over-the-air or over cable as 1080i and 1920×1080 resolution interlaced.

DOI: 10.4324/9781003247036-4

What is interlaced? In the decades prior to HD, television imagery was acquired and transmitted as an alternating pattern of lines. Odd lines first, then even, 30 times a second. The advantage, again, was that less processing bandwidth was required. But also, the interlacing allowed for fast motion. As objects tracked across the screen, there were no motion artifacts or blurring. This produced the crisp, life-like images we associate with sporting events or other live programming, such as award shows.

With 1080i HD, we now had the same motion imaging with four times the resolution and 60 lines every second instead of 30. So why didn't everyone just adopt 1080i from the start? For the production companies, they wanted to have progressive archives of their shows which were still shot on film at 24 fps (frames per second). For TV stations, they still had to incorporate footage and equipment from the standard definition era, so the scaling from 480 lines to 720 was not as bad as having to scale to 1080.

Where does UHD 4K fall into all this? One definite answer is you will NOT be seeing the network broadcasters upgrading to over-the-air 4K remotes, special events, and sports like they did for HD. Most are still making payments on the upgrade they did from standard definition to HD back in the late 1990s. Filmmakers have been using 4K-5K cameras for acquisition and then down-converting to HD/2K for presentation.

On the consumer side, there was a large push to get 4K televisions into homes. The problem was that there was no 4K content readily available to watch! You had to have high throughput internet for 4K streaming or a dedicated UHD 4K Blu-ray player. There are a few markets now where 4K is being broadcast over the air in response to televisions being outfitted with newer ATSC 3.0 tuners, but the content may still be upconverted HD programming.

The Players

In the world of live event video production, there are two main bodies that oversee standards: SMPTE and VESA. SMPTE (smpte.org) stands for the Society of Motion Picture and Television Engineers. This group outlines the specifications for formats, transmission, and processes for the production of film and television content. The joy of working with SMPTE

Table 4.1 Common serial digital formats and their associated SMPTE standard.

Resolution	Format	Data Rate	SMPTE Standard
1080i, 720p @ 59.94	HD-SDI	1.5 GB/sec	SMPTE 292 M
1080p @ 59.94	3G-SDI	3 GB/sec	SMPTE 424 M
2160p @ 59.94	12G-SDI	12 GB/sec	SMPTE ST-2082
8K and beyond	Ethernet	100 GB/sec	SMPTE ST-2110

standards are that they do not change very often and the broadcast equipment that is built to support these formats usually can accept multiple formats without issue.

Connectors and cables are usually tied to a standard as well. For instance, SDI (serial digital interface) is exclusively used with SMPTE formats.

VESA, which stands for the Video Electronics Standards Association, is tied to the Consumer Electronics Association and is responsible for standards related to computer displays and consumer televisions. Every January in Las Vegas, except during the pandemic of 2020, the CES (Consumer Electronics Show) descends upon the city and every manufacturer big and small, comes out to hawk their latest gadgets.

Unlike SMPTE formats, VESA is constantly adding new formats. This has a great deal to do with the computer graphic card and display manufacturers. Demands for higher resolutions and refresh rates by consumers creates the influx for new standards. However, certain technologies require cooperation between the two standards organizations. One example, HDR (high dynamic range), requires both SMPTE standards for acquisition and encoding, and VESA standards for proper decoding on HDR capable displays. By the way, we mentioned in the previous section about the lack of 4K content, but have you tried to find a 40-inch or bigger television in stores these days that wasn't 4K? You can thank VESA/CEA for that.

Stuff You Should Know

OK, folks, this next section is going to dive into the gritty details of what makes up a video signal. You may be surprised as what goes into the picture on your television or computer screen.

Pixels

Pixels are to video as cells are to biology. Everything is made up of them. Horizontal and vertical pixel counts make up a resolution. Resolutions can fall into one of two categories, standardized and custom.

Pixel Clock

The frequency at with pixels are transmitted expressed in hertz. Equipment and cables can have a specification for the maximum pixel clock they can pass or process. The quick formula is:

Horizontal Resolution × Vertical Resolution ×
     Refresh Rate = Pixel Clock (in Hz)

Ex: 1920 × 1080 × 59.94 = 124291584 Hz = 124.29 MHz

The more precise calculation involves including the blanking areas of the signal, the area outside the visual canvas. So, for HD 1920 × 1080:

2200 × 1135 × 59.94 = 149670180 Hz = 149.67 MHz

Bandwidth

Bandwidth refers to the amount of data that can be transmitted on a cable or network expressed in bits per second. Some common bandwidths are:

- Typical home internet connection   10 Mbps
- USB 2.0   480 Mbps
- HD 720p over SDI   970 Mbps
- Gigabit Ethernet Switch   1 Gbps
- HD 1080p over SDI   3 Gbps
- USB 3.0   5 Gbps
- UHD 4K over SDI   12 Gbps
- HDMI 2.0   18 Gbps
- HDMI 2.1   48 Gbps

Progressive

Ninety-nine percent of current video devices are progressive in nature. The little "p" next to formats, such as 720p and 1080p, means that full frames of video are displayed in progressive succession versus drawing every other field as with interlacing. The number of frames per second can vary depending on format. 1080p @ 59.94 displays 60 images per second while 1080p @ 24 displays 24 per second.

Bit Depth

This value is expressed as a bit number such as 8-bit, 10-bit, and 12-bit, and refers to the number of colors a display can produce. 8-bit has 256 values each for red, green, and blue, respectively, for a total of 16,777,216 colors. Meanwhile, 12-bit has 4096 values for each of the primary colors making for over 60 billion possible colors.

Your first instinct may be to use the highest value of bit depth to get the maximum number of colors. This might be the case for acquisition or archiving of images. However, in real live event applications, higher bit depths require higher bandwidths. Often the technology you're using has a maximum bit depth of 10 bits, so you're forced to use less colors however to most audience viewers they usually can't tell the difference between 8-, 10-, and 12-bit content.

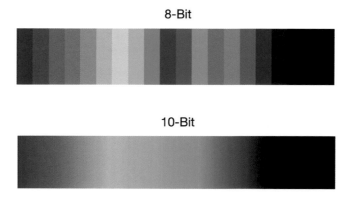

Figure 4.1 Visual comparison of varying bit depths.

Color Space

Color space refers to how the red, green, and blue components of a signal are handled. In professional video, there are two types of color spaces that are used. There is SMPTE color space and there is RGB color space.

As it was mentioned earlier, SMPTE is the organization that standardizes broadcast formats. They process color using what is called color difference signals or Y, R minus Y, and B minus Y, where Y is the value for luminance. The red and blue channels carry color information but have no luminance information, while the luminance channel carries all the brightness and contrast information as well as the green color information.

In RGB, not only does each channel carry the color information, but each color channel has a luminance value.

In SMPTE color space, if you were to lose the Y channel of the signal, you would have no picture regardless of having the other two channels. In RGB, if you lose one of the color component channels, you still have an image, just missing that one color. To dive in a bit more technically, the Y

Figure 4.2 Example of Y, R minus Y, B minus Y channels. Red and blue difference channels contain no luminance information.

Red   Green   Blue

Figure 4.3 Example of RGB channels. The brighter the luminance level, the more saturation of that color in a given portion of the image.

channel of a SMPTE signal also contains the sync timing pulse required to produce images. Whereas the RGB signal has a sync timing signal on each of the individual color channels.

Suffice it to say that RGB has much better color reproduction than SMPTE. So why not just use RGB for everything? Again, the answer comes down to bandwidth. Picture this example. Let's say you are doing a color by numbers painting. You have the outline of what you want to paint already on the canvas including all the green parts of the image. To complete the picture, you simply add in the red and blue colors needed to make the painting complete. Now imagine that you have three canvases. All three canvases contain an outline of the image you want to paint. On your first canvas you paint all the red parts of the image, on your second canvas you paint all the green parts of the image, and on your third canvas you paint all the blue. Which do you think is easier to transport from place to place, one canvas or three?

Transmission throughout a television station or through the airwaves is something that the SMPTE engineers had to consider. The component color difference color space was how they were able to reduce bandwidth with regard to color information.

For computer users, the graphics display card was usually not far from the display it would be attached to, and the signals did not have to be transmitted over long distances or airwaves. This meant they could offer the user the fullest possible color range with RGB. Even today, computer signals that need to be used on-air in broadcast must be converted to a SMPTE compatible format before they can be used.

Chroma Subsampling

Chroma subsampling refers to the number of times each color component is sampled per pixel. Chroma subsampling is often displayed as a ratio, such as 4:4:4, 4:2:2 or 4:2:0. Let's start with the RGB color space we mentioned earlier. Each color channel has a chrominance and luminance component. The minimum number of samples required is two; one and a backup. So, for the red channel of an RGB signal we have two samples for the luminance or brightness and two samples for the value of red or chrominance. This gives the red channel a value of four samples and repeating that for the green channel and blue channel gives you a total chroma subsampling of 4:4:4.

For the SMPTE color space, it is typical to use a 4:2:2 color subsampling. Again, we start with a Y channel, or green channel, which is the only channel containing luminance. Two samples for luminance and two samples for chrominance give this first channel a value of four samples. Since the color difference channels don't contain luminance information, they only require two samples for chrominance only, so the red and blue channels only get two samples each, and therefore the ratio 4:2:2.

In 4:2:0 color subsampling, even less color information is acquired from the pixels. Again, every pixel is sampled for luminance and green information giving the first value of four. However, for color sampling, only the blue information is sampled on every other pixel on one line and the red sampled on every other pixel on another line of video.

At this point, the question you might be asking yourself is, "How is the color determined for the pixels that are not sampled for color?"

The color is determined by interpolating the values of the other surrounding pixels. This pixel color "guessing" happens sixty times a second and is again done to save bandwidth for transmission. Most viewers are rarely exposed to true 4:2:2 images. You may find it surprising to learn,

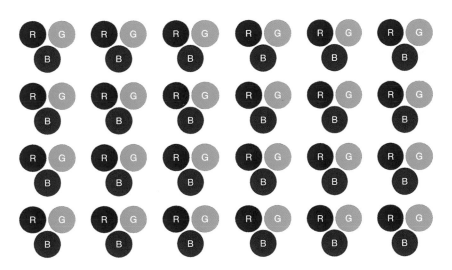

Figure 4.4 Example of RGB 4:4:4 sampling.

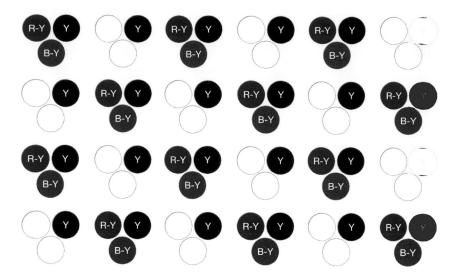

Figure 4.5 Example of SMPTE 4:2:2 sampling.

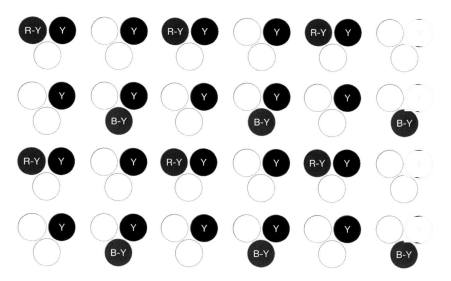

Figure 4.6 Example of SMPTE 4:2:0 sampling

but the 4:2:0 color subsampling is the most common, used for 4K streaming encoding and UHD 4K Blu-ray.

ACES

The Academy Color Encoding System was developed by industry technologists to unify the color space designation and standards for film production. With the introduction of digital technology into the film-making workflow, there needed to be a way to harmonize color management from digital cameras and film scans.

Through the use of input and output transforms based on lookup tables (LUTs), color space information is transferred from digitally scanned film and cameras in a consistent RGB format (SMPTE 2065–1) throughout the production workflow. It also supports high dynamic range (HDR) and wide color gamut (WCG).

In the live events field, this may be relevant if you find yourself on a virtual production stage managing a large LED volume. Also, content creators may be using ACES for color space management instead of traditional RGB.

Brightness

Contrary to what you might think, brightness is the term used to measure the black level of an image, not how bright the image is. When using a multi-step grayscale to set up a display, you are looking at the black bars of the grayscale when adjusting brightness level. The idea is to adjust the brightness level so that you can see a separation between the darkest bar and the one next to it.

Contrast

Contrast is used to measure the white level of an image. Again, using the grayscale chart, you would be looking at the white bars of the scale when making contrast adjustments looking for separation between the 100% white bar and the one next to it.

Limited Range versus Full Range

Consumer video displays use 8-bit values to establish a range of gray-scale from 0 through 255. They typically use what is called limited range. Anything in the image that has a brightness level of 16 and below is treated the same as pure black. Any contrast level that is above 235 is treated as pure white.

This is different than a computer monitor, which typically has full range and can display the full grayscale from values 0 through 255. Showing content that has full range values on a limited range display will cause the dark areas to be crushed and the lighter areas to appear washed out. This typically becomes problematic in live events when using confidence displays or foldback monitors for presenters on stage. Content generated from a computer, such as PowerPoint slides, when shown on a consumer monitor may not have the full grayscale range and may not appear correct. Many image processing systems have a way to set full or limited range on their outputs. If the output is set for limited range and the input coming in is a full range signal, it will be properly converted and the image at the output will display properly. Most professional series monitors have a menu item that allows you to select between full and limited range settings. This is not usually the case with consumer televisions.

EDID

Extended display identification data is a way for a display to communicate its capabilities to a source. It allows for auto configuration of resolution, color space, and bit depth. EDID emulators are often placed on the output of devices to force a particular resolution no matter what display is attached. This can be helpful in live events when you want to make sure a device is always putting out a particular resolution and not switching between resolutions as it being connected or disconnected.

HDCP

High-bandwidth digital content protection. Here is how this concept was "supposed" to work. When the first Blu-ray players were being developed, the content owners (i.e., major movie studios) wanted to have a way to ensure that their intellectual property could not be easily copied or duplicated since the picture quality of HD Blu-rays was on par with what viewers could see in the movie theater.

The solution was HDCP. The content creators would program a flag on their disc that when read by the players would force them to check for HDCP compliance from all the displays connected to the player. If any of the displays were found to be noncompliant, the entire output chain was shut down. This would allow unprotected content, such as homemade movies or corporate produced media, to be played without issue, regardless of the type of displays attached.

But that's not what happened. The Blu-ray player manufacturers (JVC, Sony, Panasonic, etc.), I'm sure in an effort to protect themselves from liability, made sure that their players invoked HDCP compliance all the time. This was exacerbated by the "analog sunset," a movement in the early 2000s to remove all analog over-the-air signals and analog outputs from consumer video devices. For consumers, this was generally a non-issue, as they were starting to buy flat screen monitors with HDMI inputs to go with their HD Blu-ray players. For live events and the professional AV market, this became a nightmare.

The trick became how to interface Blu-ray players into existing corporate AV structures that may be using older analog display equipment or require routing to several destinations. For live events, how do you extend an HDMI signal off a laptop or Blu-ray player that wasn't designed to go more than 15 feet?

Today, there is no technical reason why you cannot have a fully HDCP compliant signal chain from source through switching and out to displays. There are signal transmission options that are compliant for both video over Ethernet and over fiber. The number of "keys" allowed per device has been increased, so a single HDCP source can be routed to multiple HDCP destinations.

But just when you thought you had all the HDCP solutions, it makes itself more complicated. With the introduction of HDCP 2.2 for 4K UHD content encryption, there is now another pothole in the video signal road to lookout for. At least HDCP 2.2 has two modes. The first prevents all from being displayed if there is no encryption. The second allows HD versions of the content to be played only if there is no HDCP 2.2 encryption. So, if you bought a 4K television several years ago before HDCP 2.2 was released and you're wondering why you can't get any 4K streaming content, now you know why. There is no way to upgrade the HDCP in your older 4K television to the newer version.

Genlock

This is using an analog sync pulse, such as black or tri-level, to synchronize the components in a video system. By running a centralized sync pulse to all genlockable equipment, you ensure that the rendering of pixel 0,0 happens simultaneously across all devices. This makes clean switching during the vertical interval possible.

In today's digital age, asynchronous signals can be fed to video switchers without being genlocked. The switcher will frame buffer the video for one frame while it tries to synchronize with the internal sync of the switcher. Although it is just one frame, the aggregate of frame delay through a system can be visually significant if all devices in the processing chain are having to add a buffering frame.

This is not the same as frame locking.

Frame Locking

When using multiple outputs of a graphics display card to create large backgrounds or canvasses, or to synchronize frame creation between two or more separate computers, use frame locking. This can be a separate Ethernet connection between units or timing information sent over the shared network. It ensures that all GPUs (graphical processing units) are rendering the same frame at the same time. However, without the additional genlock signal, they could be on the same frame, but at different lines in the render. This would cause tearing in the seams where the content is edge-butted together on a display, such as an LED wall.

Connector Types

HDMI

High-definition multimedia interface. This connector became the default standard as consumers made the transition to high definition. A single cable would allow for digital audio and video to be transferred in high definition with 5.1 surround sound. This connector would find its way on to consumer laptops, replacing the analog DB15 pin VGA connector.

We are currently on HDMI version 2.1, a standard that supports 8K video signals. This chart shows the previous versions and their resolution limits.

The inherent problem that comes with HDMI, and this is somewhat typical of all connectors, is the ability to identify what version a cable may support. When HDMI was first released there obviously was no reason to distinguish the cable in any way shape or form. With the onset of HDMI 2.0 now users had an issue of how to identify a cable that supported this new standard. In the live event and AV industries, problems were attributed to gear being faulty, when in fact it was the improper version of HDMI cable that was to blame for the issue.

With HDMI 2.0, if a user found a cable that had the jacket printed with the words "high speed" or "HDMI plus Ethernet," then there was a good chance that the cable could pass a 4K UHD signal over short distance. Unless the headshell or cable jacket is specifically marked with an HDMI version, or some shop techs have taken the time to test the cables and label them appropriately, you should assume the cable is an HDMI version 1.4 and you should find another cable for specific 4K uses.

Also be aware of the difference between specifications and implementation. All manufacturers using HDMI ports on their products pay

Table 4.2 HDMI versions and their associated rates and maximum resolutions.

HDMI Version	Max Transmission Rate	Resolution Refresh Rate
HDMI 1.4b	10.2 Gb/sec	4K @ 30
HDMI 2.0	18.0 Gb/sec	4K @ 60
HDMI 2.1	48.0 Gb/sec	8K @ 60

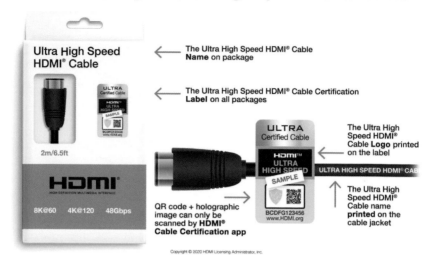

Figure 4.7 The HDMI Forum is making attempts to work with cable manufacturers in the proper identification of ultra-high speed cabling for HDMI 2.1. Courtesy HDMI Licensing Administrator, Inc.

a licensing fee. However, use of the HDMI port does not mean that the manufacturer must include all the features of a particular specification. Only what they decide to implement must meet the specification. In other words, just because there are HDR specifications for HDMI 2.1 and the manufacturer lists their product as HDMI 2.1 compliant, doesn't mean that their device supports HDR unless they specifically say so.

For running HDMI signals over distance, there are hybrid cables which have a send (source) and receive (monitor) end. They must be run in the correct direction. In the source headshell are active electronics that are powered by the five volts supplied by the HDMI port. This is a fiber optic transmitter which is run alongside a small copper wire to deliver voltage to the fiber receiver at the other end.

DisplayPort

Like HDMI, DisplayPort (DP) is a connector meant to replace another legacy type: DVI. There are four lanes linked to provide 10.8 Gbps bandwidth.

In the 1.1 specification, 4K resolutions are supported up to 30 Hz refresh rate and up to 60 Hz in the 1.2 specification. Additionally, version 1.2 increases the bandwidth to 17.28 Gbps.

Unlike HDMI, DisplayPort does not have direct pin-to-pin compatibility with DVI. In fact, most conversions from DisplayPort to anything else require an active (USB or bus-powered) converter. DP converters are directional and are typically not marked as to which direction the conversion is happening. When grabbing video adapters for use with DisplayPort, if they do not work, it may be that they convert in the opposite direction than what you are trying to do. Personally, I have had much success with DisplayPort adapters from Club 3D (www.club-3d.com).

DisplayPort has been used under the name Thunderbolt or mini-DisplayPort on Mac computers for some time and is seeing use on PCs as USB-C ports. Both USB-C and Thunderbolt are data busses that can transfer data, such as to an Ethernet interface, or provide a video output from the onboard graphics processor. DisplayPort simply refers to the video output portion of the specification.

It may be the case that on computers with multiple USB-C connectors only one is designated for video output. Especially if outputting 4k resolutions. Computers that support DP 1.2a may support MST (multi-stream transport), which means the video signal can be daisy-chained through displays that support MST and multiple desktops with different resolutions are supported. DisplayPort ++ is another option that allows one DP port to be split into up into three discrete video feeds via a hub.

SDI

This is serial digital interface. In the real world, SDI cable may go by the generic name BNC cable. Named for the type of connector used at the ends, BNC cable should be inspected before use to ensure it is rated for the bandwidth of video signal. Most SDI cable will have the term "digital" printed on the jacket along with a Gigahertz rating: 3GHz for HD 1080p @ 60 Hz or 12 GHz for UHD 4K @ 60 Hz. If neither of those ratings are visible or the jacket has no markings, it may be a legacy analog cable mixed into the pile.

SDI follows SMPTE standards for signal format. Up to eight channel pairs of AES (digital audio) are included in the specification for the SDI. By embedding audio with the video signal, you can make a single connection for SDI capable recorders or transmit hundreds of feet to another location where you can unembed the audio from the video signal. There have been instances where audio embedder/de-embedders and SDI cable were used as a multi-channel audio snake in a pinch.

As with all of these digital signal cables, the ability to maintain a viable signal is based on an analog voltage-based clocking signal. The clock signal maintains the order of the ones and zeros in the digital stream. When this clock signal becomes weak, the digital picture falls apart. Likewise, the amount of data being transmitted is inversely proportional to the distance it can travel.

Video cable gauge is organized by an RG rating. RG-6 cables are the size your local cable company uses to wire your home. Any long SDI that run 100 feet and longer use an RG-6-gauge cable. RG-59 is a bit smaller in gauge and may be used to go 25- to 50-foot distances. Mini-coax is usually reserved for internal wiring inside racks.

Cable specifications give optimum distances for various signal formats. A cable rated for a maximum distance of 250 feet when using a 12G 4K signal, could probably run an HD 1080p signal 500 feet; less data, greater distance. The quality of the SDI transmitters and receivers in various equipment can also affect transmission distance. For manageability, most production companies do not go beyond 250 feet for the length of their copper cables. Should you need to go a father distance, you can place a re-clocking device such as a reclocking DA (distribution amplifier) inline to restore the clock signal levels and drive the signal another 200 feet.

Fiber Optics

When transmitting signals over long distance, most companies go straight to fiber optics instead of using long copper cables. Once a very expensive alternative, affordable fiber-optic transmitters and receivers, and the low cost of military-spec fiber cable, have all made fiber a viable solution.

The transmission of signals using wavelengths of light allows for distances of hundreds of miles. Fiber is chosen because there is no degradation of the signal between the origin and the destination. It can transport video, audio, data, or all simultaneously.

Since the material at the core of this technology is essentially glass, care is taken to protect the core with thick jacketing (sometimes using Kevlar) and to limit bending the cable past a certain angle.

There are two types of fiber modes: single mode and multi-mode. The light source for single mode is a laser or laser diode. It has a 9-micron optical core, which is half the diameter of the average human hair. Used for extreme distance applications, single mode can cover distances that are dozens of miles. With higher bandwidth capabilities and the use of actual lasers, single mode is generally more expensive than multi-mode.

Multi-mode uses LED as its light source and has a shorter distance limit at about ten miles. The optical core is bigger at 50–100 microns in size. Multiple cores are used to increase bandwidth. The OM rating (OM 1,2,3,4,5) designates the bandwidth capabilities with a 10-fold increase between ratings. OM3 is the most common type for data applications, and it has a capacity of 10Gbps and an effective distance of 300 meters (990 feet). The newest rating, OM5, is mainly for use with data transport of 10Gbps and above and extending the distance of 10Gbps to 550 meters (1800 feet).

There is no interoperability between single mode and multi-mode fiber. You can adapt from one connector type to another with in the same mode or barrel between connectors to extend a cable. Here are some common fiber-optic connectors.

The jacket color of a fiber-optic cable will also indicate its type and bandwidth.

The long runs used in live events are multiple pairs of fiber cable and will be covered in a black outer jacket. Hopefully the cable type will be printed on the jacket. The use of Neutrik opticalCON connectors is quite prevalent. These cables contain multiple pairs of fiber cables. Then, using different fan-outs, they can quickly switch between connect types.

Dust is a fiber-optic cable's worst enemy. Caps are often supplied to protect the ends of the cable during transport and handling. Do not remove the dust caps until you are ready to make a connection to a piece of equipment. Make sure to replace the caps on any unused fiber connections and when packing the cable for transport. A good item to have in your toolkit

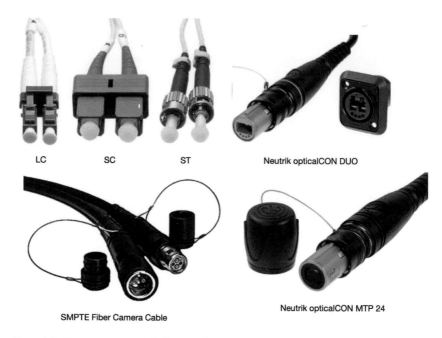

LC   SC   ST   Neutrik opticalCON DUO

SMPTE Fiber Camera Cable

Neutrik opticalCON MTP 24

Figure 4.8 Common types of fiber-optic connectors.

Table 4.3 Jacket color meanings of fiber cables.

Jacket Color	Type
Yellow	Single Mode Fiber
Orange	Multi-mode Fiber
Aqua	MM OM3,4
Lime green	MM OM5

is a fiber cleaning pen. The end slides over the glass tip of a connector and a slight pushing action, along with a haptic click, cleans the optical connector. This is an easy fix for cables that are dropping signal or not showing signal at all. Dirt on the tip can scatter or prevent the light source from propagating correctly down the light pipe.

Another tool is a visual fault indicator. This is a light source, similar to a red laser pointer, that attaches to fiber cable at one end. When inspecting the jacket, should you see a red light somewhere inside the cable, that indicates a break in the glass fiber core. If anything, it helps you confirm a questionable cable. However, with the black jacketed cables, this is some-what useless as the light dot becomes impossible to see.

Troubleshooting Cables

When things start to go wrong with signals and images, the first thing that happens is trying to find fault with the most expensive pieces of gear. If a projector blinks, there must be an issue with the projector. If a laptop feed drops out, there must be a problem with the switching system. In many circumstances, it is the interconnecting cables that are at fault.

Did you use a HDMI cable compatible with the version that allows 4K @ 60 Hz? Or did you just grab the cable from a bin of unmarked HDMI cables? Is that adapter a DisplayPort to HDMI or the reverse?

HDCP encryption can also be cause of frustration. Many devices have HDCP invoked or enabled by default. If the final output is going to be SDI, which doesn't support HDCP, then there is no image. Make sure your input and output settings have HDCP turned if there is no need for it on your event.

There used to be a saying, "cables don't go bad." The fact is they do. They can fall out of specification or be used past the recommended num-ber of insertions. Cables with active electronics built into the headshell could get damaged or shorted out. Never dismiss the stuff in the middle.

5

THE BIG PICTURE

It would seem logical that live event video would have a basis in broadcast technology. After all, large screen displays at sporting events and concerts showing content from broadcast switching systems is commonplace. But video systems for a live event, such as a major award show or large corporate meeting, are as different from creating local broadcast news programs as recording studio production for an album is to mixing live sound in a 50,000-person arena.

The saying among AV and live events people is that you don't go looking for this career, it finds you. A common story is a college student answers an ad for "hotel AV technician" from the college career center or online career site. Being in a band, they have some rudimentary experience with sound systems, microphones, and a little lighting. They pass the interview and are taken on a tour of the hotel's ballrooms. For the next three to five years they spend weekends setting up sound systems and lighting trees

DOI: 10.4324/9781003247036-5

for weddings and bar mitzvahs. During the week, they are setting up flip-charts and small video projectors for business meetings.

Then a large corporate show loads into one of the ballrooms. They see equipment they have never seen before: large format projectors, flown speaker clusters, and moving lights. Gear normally seen on the biggest rock tours and Broadway productions. Curiosity forces them to find a way to befriend a member of the production crew. They ask who they work for and where can they learn to work with this equipment? With a little research, they learn the names of two or three rental staging companies in the area and submit applications.

Another path is a person who works at one of these rental staging companies, tells a friend they have openings for warehouse people. The job is just checking gear in and out and truck loading. A few years go by, and the friend has learned to operate the equipment she is handling in the moments of downtime she has each day. She moves into a quality control position and eventually goes out as a technician on shows.

Both these stories are ways of paying your dues. The classic system of working from the bottom up. You'll spend years pushing boxes before getting to sit in the hot seat and operate an event. Patience is key and can lead to a life of wonderful memories with lifelong friends and unique locations.

Formal education in live event technology has been sparce to nonexistent. For decades, converts came with degrees in broadcast engineering or theater arts. People coming from the world of television were blindsided by having to interface with computers and deal with the various non-standardized signal types. Audio mavens pivoted from concert touring to long days keeping tabs on wireless microphones and having to cut their hair.

Training was mostly on the job, and in the case of video, it was not uncommon to see a piece of gear for the first time on a show site, and you had until the first rehearsals to figure out how it worked. Before high-resolution switching systems, broadcast production switchers were utilized, and this is where you had to bring someone in from the broadcast world. This was because the station or network they worked for paid for their training. It was not unheard of for Grass Valley Switcher training to last a full week and cost $10,000. This cost was usually absorbed by the networks or stations as part of the purchase of the equipment. The manufacturers saw that their equipment was being used in non-traditional ways and opened training to freelancers at little to no cost. The more people that knew your equipment, the more it would get specified and used.

In today's world of YouTube and e-learning, there's a treasure trove of instructional videos and manufacturer-created QuickStart guides to help get through most of the new equipment put out these days. Offline and simulation programs will let a technician play with the software interface for a piece of gear, such as a media server, and learn the order of operations and features without having to have the actual unit present. The downside to all this over information and accessibility is overconfidence. It's one thing to run a piece of gear in a simulator and another to actually use it on a show site and have to contend with all the different signal cables and pieces of equipment you need to interface with. It's all about baby steps. Start with small events and configurations that don't require all of the gear's full feature sets or capabilities. Be comfortable in your knowledge levels and don't take on shows that are beyond your abilities. You're only as good as your last show.

But as a video technician, are live events the only avenue I have to use my talents? The answer is absolutely not. The world of audiovisual and entertainment technology spans many verticals that your talents could be used for. It may seem counter intuitive for a book about live events technicians to cover other job possibilities. But given recent events, to know where your skills could also be applied, should you need to pivot careers, may be helpful.

It may seem obvious that a person working in live events video could find work in the broadcast industries. In particular, working at a television station or network. Unless you began your career in broadcast and made the transition to live events, going the other way is much more difficult. Broadcasting is a regulated industry requiring licenses and certifications for most personnel operating a television facility. For broadcast engineers this means FCC licenses and a minimum of two years formal training in broadcast technology. For other operating personnel in a television station, such as master control or camera operators, SBE (Society of Broadcast

Engineers) certifies audio and video operators. If you worked for some time in the live events industry, you probably have sufficient knowledge to be an assistant engineer, camera operator, or control room operator. Be aware that more broadcast facilities are turning to automation in order to reduce staffing and the positions that do remain may not be as financially rewarding as live events.

Another vertical is fixed installation. This is where video systems and displays are installed permanently in corporate lobbies, meeting rooms, or command and control rooms for utilities or crisis management. Basically, AV that doesn't move.

The advantage here is that projects are usually planned out months to years in advance. Entire video systems are designed, fabricated, and wired offsite in the months prior to the actual installation. Once on site, all final connections are made, and the system is tested. If you like a sense of permanence to your work and slightly lower stress levels, then working for an integration company or as a consultant in the fixed installation world might be for you.

An offshoot of the fixed installation vertical is amusement parks and themed entertainment. A feature of many of the world's major amusement parks is some form of projection mapping. Using multiple high lumen projectors aligned to architectural structures, preproduced content is displayed in conjunction with an audio soundtrack and lighting effects. Projectors may also be used with water effects or as display elements within attractions themselves. The more creative person may take the time to develop skills in content creation and video editing.

Finally, houses of worship may be an option. Churches of all sizes have the need for web streaming or live video production of their services. Today's mega churches will have HD broadcast camera switching systems, multiple cameras including jibs and cranes, and full audio and lighting rigs to rival any theater. If you're looking to explore a career in live event video, or audio and lighting for that matter, maybe consider visiting a church in the area with these production capabilities and volunteering to assist on Sundays. You may find it a very cost-effective technical education.

Industry Associations

Several organizations exist they can help you with networking, career advice, job listings, and many have annual conventions where you can walk trade show floors and see the latest hardware from industry manufacturers. There are learning sessions which can cover various subjects as well as equipment training.

One of these is a AVIXA (www.AVIXA.org). The Audio Visual and Integrated eXperience Association holds an annual conference called Infocomm. Historically focused on the fixed installation market, AVIXA has opened up to include the live event community. They offer the CTS certification which is an industry-recognized qualification of technical knowledge. Variants of the CTS certification are CTS-I which focuses on integration related specialties and CTS-D for skills related to the design of AV systems. AVIXA offers scholarships for students studying AV related subjects and has learning resources related to the industry.

LDI (Live Design International) takes place yearly in Las Vegas and brings together companies and people related to live events technology and execution. The trade show floor is usually a sensory overload of lighting, effects, and other hardware related to production. Companies featuring projection and LED display products will also exhibit at this conference.

The National Association of Broadcasters (www.nab.org) also holds its annual conference and trade show in Las Vegas. This is the convergence of all things related to video (and film) production. From cameras and production switchers, to fiber transmission and signal processing, NAB is the premier event for anyone involved with video at any level.

Computer Convergence

It has come to the point where you may need a computer science degree to be a video engineer. The convergence of computer and video technology means that most video devices run on embedded OS systems or on highly modified server platforms. All of it is networked on multiple VLANs requiring managed switches and IP address allocation. At the very least, a CompTIA+ or Google IT Support certification could fill in the networking and computer hardware gaps.

There was a time when control surfaces were all buttons. Then younger operators asked for computer GUI control options that integrated with their touch devices, eventually wanting to know the code snippets so they can program their own button panels.

With the recent onset of virtual production, and by virtual production we mean the remote switching of sources to a webcast or webinar, the physical switcher has been replaced by software driven solutions such as vMix and OBS. This has done for video production what PowerPoint did for graphics production. With a good enough internet connection, a production hub can be assembled in the spare bedroom of your home.

The COVID-19 pandemic has assured that all in-person meeting events will have some element of virtuality. Any system design for live events will need to include a method of ingest for remote presenters, as well as web-streaming. But at some point, couldn't a live event with all prerecorded content not be equally served as a YouTube video?

Continuing Education

Like you heard in so many graduation speeches, learning is a lifelong process. Using downtime or seasonal slowdowns to gain extra knowledge is always a good thing. Although many resources are free, such as YouTube videos, there are some manufacturers that charge for equipment certifications. This is often the case for in-person instructional courses.

Does it make a difference to have a certification? If you are developing your client pool or want to impress the clients of a labor broker, then the answer is yes. Does a certification prove you can operate the gear in a show environment? Absolutely not. That comes with experience and building your reputation from show to show.

There are those that say certifications are useless and a waste of money. Some can argue that the most successful technicians don't have certifications for the gear they specialize in. Lucky for them that they had the opportunity to learn the gear either from exposure or the initiative to go into a company's shop and play with the gear. For the rest of us, a manufacturer's course not only gives us hands-on time with the product, but we learn how the manufacturer intended for us to use their product. This can often dispel myths and misconceptions that may circulate amongst users. At the end of the day, certifications are like the Department of Motor Vehicles. The driving test can be passed with a score of 100 or 70, but that has no indication of how good a driver you will be after getting your license.

6

SWITCHING SYSTEMS

Despite all the technology that goes into a typical television news program, such as cameras, playback devices, character generators, green screens, remote feeds, and satellite downlinks, it is all meant for a single destination: the television in your home.

Production switchers, or vision mixers as they're called in some parts of the world, can be made up of single or multiple mix effects buses, or MEs. A switcher at your local TV station may have three or four MEs on it and may look daunting when sitting behind it for the first time.

But it's really not as complicated as it may seem. Each ME is made up of three rows, a preview or preset bus, the program bus, and the key selection bus. From left to right these buttons are mapped with the sources coming into the switcher and the sources are aligned across all three buses. For example, camera 1 may be coming in on button two of the switcher, so therefore button two on the preview program and key selection bus are all camera 1. Taking this concept a step further, the next ME

DOI: 10.4324/9781003247036-6

Figure 6.1 A large three ME Snell & Wilcox Kahuna HD production switcher used for a corporate event. Photo by Tim Kuschel.

is just a duplication of the previous ME in terms of buttons. Button two on any other ME would select camera 1. So why so many MEs?

The reason is that multiple MEs allow operators to set up different effects ahead of time before sending them to their main output program. Otherwise, they would have to do all their effects from the same ME bus that they use to switch the program feed to air, and that may get confusing and cause mistakes. The setup for a weather segment using a green screen might be set up on ME 2, while on ME 3 there are two picture and picture windows set up for a segment involving a remote reporter.

Production switching follows what is called a re-entry scheme. ME buses can be fed into one another but ultimately must come out one single bus that is dedicated to the feed that goes on air. What assists the technical director in operation of the production switcher is the use of memories and macros. The memory system on a production switcher can save the settings for a single ME or multiple MEs at the same time. Those memories can be recalled via a numerical keypad and instantly reloads a particular look to the ME it was saved from. Macros are a sequence of steps that can be recorded as a single cue that when recalled executes those steps sequentially or with any real-time delays that were present when recorded.

Traditionally the memory system records everything about an ME, including sources selected on the preview bus the program bus and the key selection bus. More recent switchers have had the ability to designate which buses are recorded as part of a memory. This has been very helpful in using production switchers in other live event situations that may be cue driven where a preset might want to be reloaded but not affect what is currently on the program bus. The program bus is disabled when the memory is written so that when it's recalled there is no effect to the program bus. These type of selective conditions for memories are usually referred to as enables.

When production switchers were first being used for multi-screen production, the methodology changed. Instead of MEs feeding into each other and out a single bus to a single destination, each ME itself would be sent to a different screen destination. For example, if you had a three ME switcher, ME 1 might be feeding the house left screen, ME 2 the center screen, and ME 3 the house right screen.

The advantages to using a production switcher for multi-screen switching were many. The first main advantage was that all your sources were a single format. With production switchers, they conformed to broadcast standards and all sources had to be same format coming in. On a HD switcher set for 1080p, the camera and other broadcast quality equipment that supported HD-SDI would be an input to the switcher at the native rate of 1080p. Any other devices, such as computers that output HDMI or display port, would have to be converted to the HD-SDI format before they could be used. This concept of converting to one format is popular with a lot of video production companies as getting everything to one format helps reduce issues and troubleshooting. The other advantage to using a production switcher is the ability to use advanced effects for transitions and PIP windows, storage, playback of motion clips, and any integration with virtual studio technology. Production switchers also have what's called an aux bus. It's usually a separate row of buttons at the top of the control surface that acts as a routing matrix. Input sources and program feeds generated by the MEs can be routed to aux outputs. The number of aux outputs available is usually a function of the size of the switcher. Smaller one ME units will only have two to four auxes, while larger ME units can have eight plus.

Aux feeds can be used to send signals that are not affected by the mix effect buses. In the case of a live news broadcast, the technical director may use the aux to control the signal going to a large monitor behind the anchorperson's desk, or the feeds to a monitor wall used by the traffic and weather reporters. A common feature of recent switchers is the ability to have dissolves on aux feeds. Ross Video even coined the phrase "mini-ME" when it implemented this feature on its aux outputs.

Production switchers have had a place in all types of live events. They may be the only tool needed to support cameras and image magnification (I-mag) for concert tour or they may be a dedicated camera switching system feeding into a multi-screen management system.

Multi-Screen Management Systems

A production switcher takes a certain amount of learning curve and expertise to switch a multi-destination event successfully. Multi-screen management systems are used in live events to manage not only multiple destinations, but multiple input format signal types. Production switchers cannot natively ingest display port or resolutions other than the standard SMPTE formats such as 1280×720, 1920×1080, or 3840×2160.

The power in multi-screen management systems comes from their scalers. A scaler is an electronic device that resize pixels in order to fill a required space. For example, and output canvas maybe 4K at 3840×2160 pixels, but the input source is only 1920 x 1080 pixels. The scaler extrapolates the pixels and resizes the input source to fill the 4K canvas. The quality of the scaler is what determines how good the image looks resized. does the image turn soft in focus or are there other artifacts due to the scaling up of the image?

In general, most scalers do a fine job of scaling resolutions up. It's when they have to reduce the scaling to something smaller than the native resolution that the better scalers will stand out. If you can reduce a 1920 × 1080-pixel source by 50% in size and still be able to read the text clearly, then you have a very high-quality scaler.

If we think back to the earlier chapter where the goal was to be able to mix between video and computer sources seamlessly, screen management systems are just an over extension of that concept. Unlike production switchers, screen management systems can deal with odd or large pixel spaces made up of multiple nonstandard resolutions.

Pixel Space versus Pixel Canvas

Screen management systems usually follow one of two methodologies when it comes to the overall pixel count that the system can manage. The first method is pixel space where each device as a large pixel area that its outputs can move within. An example is a single 4K PIP window can be shifted 10,000 pixels to the right off the visual area or 10,000 pixels to the left off the visual area without affecting any other output coming from that system. When combining or linking additional units, they each bring their own pixel space and outputs can be combined to interact within the same pixel space or completely independent.

The other methodology is to create a fixed pixel canvas. This is where the canvas has a finite number of pixels that are divided up among the destinations created by the processor. Think of cutting up a cake into different sections and once you've handed out all the cake, you are out of resources. Taking that same example from before, if you were to take a 4K PIP and fly it off the right edge of one output it would suddenly appear coming in the left edge of another output. This is because the outputs are next to each other on the pixel canvas. Another thing to consider with these types of systems is that user features, like multi-viewers, are part of the pixel canvas. But pixel canvases can have their advantages. Since it is one contiguous space, sources can be dragged across multiple screens or elements flown through multiple outputs. One thing to note, however, is that program and preview for an output destination both require pixel resources weather using pixel space or a pixel canvas.

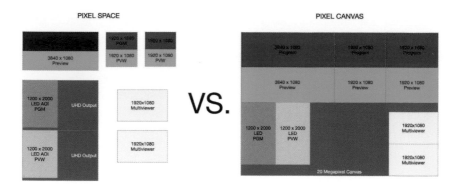

Figure 6.2 Destinations have unique pixel spaces which are independent versus the model where all destinations use an allocation of canvas pixels.

Cueing the Show

Memories are to production switchers as presets are to screen management systems. In broadcast productions, there is usually a director who calls out what he or she wants to see on program. Something like "ready camera one, take camera 1. Ready title, go title. Ready weather," etc. It is up to the TD sitting behind the switcher to figure out the fastest way of implementing what the director wants to see.

A live event is driven more like a theatrical production, with the stage manager giving standbys for cue numbers to the various departments. It's not their job to call out particular cameras or what titles to key in. This is the world that multi-screen management systems were born into, and their preset systems are customized for this type of operation. When an operator is composing a look that will be used during the event, that look is saved as it appears on the preview bus. When that preset is recalled, it will load to the preview bus only and not affect what is currently on the screens. The only time that the program screens may be affected is because the operator didn't understand that due to resource conflict, a source currently up on program was inadvertently changed by the preset they just loaded.

You'll find that operation of a screen management system is much simpler and suited for meeting environments or live cue-based events. At the time of this writing, there are three main products in this space:

Barco, with its E2 Event Manager platform, Christie Digital with their X80 Spyder, and Analog Way with their LivePremiere Aquilon system. You'll find that technology in this space tends to leapfrog. Manufacturers never release products at the same time. In fact, it could be said they wait a little bit to see what the competition puts out before finalizing their own next product.

Single Output Devices

But maybe we're getting ahead of ourselves. If you are starting out as a live event video technician, you may be relegated to controlling single screen events. The switchers used for these applications usually have six to eight inputs of various flavors, including HDMI, BNC, or DisplayPort. They have one main program output and usually a preview output for the operator to see the sources ahead of time or a multi-viewer.

Simple breakout rooms, or events where the operator is doing double duty as the audio and lighting person, are the typical scenarios in which these simple input switching devices are used. Their features may be very spartan and limited to just dissolves or cuts in between sources. One thing you should note whether it's a small single screen switcher or a large multi-screen management system, is that there is usually a memory system for saving input configurations. What this means is that once you have acquired a source on an input and made any applicable changes to it such as sizing or color adjustments, you need to save that configuration into non-volatile memory. This is accomplished by selecting "Save" on the input menu after you are done editing the source. By not doing this, you run the risk of delaying transitions during your event. Recalling saved input configurations of sources is much faster because the input channel is not having to always reacquire the signal and auto configure it every time you select it.

Another single screen device that you may find yourself using is called a scaler or format converter. These devices have the singular task of converting a signal from one connector type and format to another connector type and format. Product examples of this are the Barco-Folsom ImagePro and the Analog Way Pulse. Similar in function these products can convert a DisplayPort or HDMI signal into an SDI signal for use with a production switcher. They also allow these unsynchronous sources to be genlocked to a reference signal.

As much of a Swiss army knife as these products may seem, they can't do everything. When converting from a connector that has HDCP encryption to SDI, there will be no output because SDI does not support HDCP encryption. Extremely high resolutions that can be accomplished on DisplayPort connectors may not be scalable down to the bandwidth of another connector.

System Engineering

We should take a moment to talk about system design and engineering larger video systems. This primarily deals with managing signal flow from an input through a system to some form of output. There are best practices and ways of just getting it done.

The first thing to consider is the number and type of outputs you'll be dealing with. Let's look at a scenario with three projection screens (center, left, and right) that are projecting an HD image at 1080p. Since all the screens are projecting a standard HD format, the first thought might be to use a production switcher in this instance. If all the screens were showing the same content all the time, then this would only require a single ME switcher and a means to distribute the program feed to all three screens. However, if the request is to be able to show different sources on the three screens at any time or what is normally referred to as "anything anywhere," then you now require a production switcher that can give you three discrete outputs, which now means you need a three ME switcher or a switcher with at least two mixing aux outputs.

The other thing to consider in this scenario is that since you are using a broadcast standard, all equipment that is non-broadcast, such as computers and possibly remote feeds, will have to be converted into HD-SDI.

Now let's look at this scenario using a multi-screen management system. Any of the three major systems I mentioned earlier would be able to handle three screens with no problem. They could all have the same content or different content, and the cues could easily be set up as presets. Computers could be ingested using their native connector type, and the signal to the projectors could still be at HD-SDI.

In the multi-screen management system, you can reduce the number of converters needed on inputs which means also reducing the number of points of failure. Points of failure is a concept that you should keep in mind when designing any video system. It means the steps in the chain or places where the signal goes in and out before moving onto another device. In the example of using a production switcher, a computer's HDMI signal would have to go through a converter, changing its signal type to HD-SDI. Then from the converter it would be connected to the switcher.

If you were to lose the computer signal, you would have to troubleshoot the output of the original source and the output of the converter. Two points of failure. But being able to go out of the computer HDMI into the screen management switcher directly, you can eliminate the converter or one point of failure. But a point of failure isn't just a piece of hardware, it can be a cable in between two pieces of equipment.

As HDMI and DisplayPort capabilities increase, they can require different specifications from the cables that are used to interconnect devices. In a previous chapter we mentioned how going from HDMI 1.4 to HDMI 2.0 required an increase in bandwidth and that most cables we're not labeled with the version of HDMI that they supported. Therefore, when a technician grabs a cable to use with a 4K @ 60 Hz signal and has issues with the image, the first solution is to find fault in the hardware. It may be as simple as replacing the cable with another one labeled as rated for HDMI 2.0.

Another aspect to system engineering is knowing the limits of your signal cables. An HDMI cable with a 1080p signal can go about 50 feet. Likewise, a DisplayPort cable is rated for a maximum distance of 25 feet. Cable distance is directly proportional to the data rate and bandwidth of the signal. To put it simply, the more pixels you're trying to transport from one place to the other, the shorter the usable cable distance. A HD-SDI cable rated for 3 GHz can send a 1080p signal up to 250 feet. That same cable with a smaller bandwidth 720p signal can go 500 to 600 feet.

The standard operating procedure with most staging companies these days is to run fiber for any signal that is more than 200 feet away. In the past that would seem like a very expensive way of doing things. However, as costs have come down on tactical fiber and the end points used to transmit and receive, this is a much more viable and reliable method of transmission. Copper BNC cable may still be used backstage where projectors may be within 200 feet of video village.

Matrix Routers and Distribution Amplifiers

Since most devices only have one output connector, it may be necessary to introduce equipment that allows for routing and distribution of signals to more destinations than your switching system is capable of. This is where matrix routers and distribution amplifiers (DA) come into play.

When connecting a computer into a switching system it may be necessary to have a local monitor at the computer for the operator to see output as well as taking a signal to a switching system. By placing a DA on the computer's output, you can split the signal to two or more different places. The purpose of a distribution amplifier is to not only split the signal but to boost and equalize as well. In the digital world there are no passive "Y" cables that can be used to split the signal.

Matrix routers usually come in some configuration of inputs and outputs, 16 × 16, 32 × 32, 48 × 48, etc. Like a DA, an input coming into the matrix can be routed to multiple outputs. Or a series of inputs can be routed to various outputs. Matrix routers introduce more flexibility into the control of signal flow and are a standard element in most system designs. Another advantage to matrix routers is that the source to a destination can be changed with a few button presses as opposed to having to disconnect from a source computer and reconnect to another one physically. Through the use of salvos, routers can completely change their input and output routing assignments. These salvos can be programmed to only affect certain inputs and outputs.

Matrix routers upstream (before) or downstream (after) of a switching system takes workload off of the switching system for those signals which tend to be "set and forget" or can expand the number of inputs available to a system buy sub routing signals as needed into the available inputs.

As with cables and connector types, you need to make sure that the routing matrix or distribution amplifiers that you use match the specification of the signal you are trying to distribute. An HDMI routing matrix that only supports HDMI 1.4 would be useless if trying to run a show with 4K @ 60 HDMI 2.0 sources.

Managing HDCP

When high-definition content protection was first introduced, it caused havoc in the AV industry because most systems still were mostly analog. Add to that the fact that HDCP was invoked even on content that was not commercially produced, and now you had integrators and video technicians scrambling to put Blu-ray and mobile handheld devices on screens.

Today we have moved to a primarily digital infrastructure. There is no reason why you can't have an HDCP compliant signal path from source through switching and all the way to your projectors or display devices.

The general rule is any piece of equipment with an HDMI connector on it has paid a licensing fee to be able to use that connector and likewise follows the rules regarding HDCP encryption.

The only digital format that does not support HDCP encryption is SDI. This applies to HD-SDI, 3G-SDI, and 12G-SDI.

Also, just because a piece of equipment says it supports HDCP, be sure to do your due diligence and check the specification to make sure it matches the version of HDCP that you need for your signal types. There are very few instances now where HDCP is a major issue. Just be sure to check with your presenters on what type of content they will be presenting and if any of it comes from streaming services that would cause their devices to require HDCP connections.

Be aware that on some switching systems HDCP support is turned on by default, likewise some devices it is turned off by default. So, if you aren't getting a signal on an input or on the output of a device, make sure the HDCP is set correctly for your situation.

Genlock and Timing

With video's digital age, we entered a realm of more plug and play ease of use. Asynchronous sources can be plugged into digital production switchers and circuits in the switcher will timebase correct or resynchronize those sources to the switchers internal sync. It makes for a wonderfully easy setup and you can quickly get your show up and running. Here is where we need to mention an electronic circuit called a framebuffer. The job of the framebuffer is to hold video temporarily while it tries to synchronize with the internal sync. This can cause a delay of up to one frame. There can also be processing delays in other equipment down the line including fiber transmitters/receivers and LED pixel processors. A total video system delay can be three to five frames. The human brain can resynchronize up to three frames of delay between visual and auditory sources. More than that and we experience lip-sync issues.

Genlock is an analog pulse signal that acts like a heartbeat. The pulse signals to a video device when to render pixel 0,0. When all devices are rendering in time together, they are said to be "genlocked," short for generator locking. If all the external video devices are receiving a genlock signal, and the switcher is receiving the same genlock signal, then they can be connected and bypass any frame buffering, thus eliminating this initial frame of delay.

Sending the genlock sync pulse to other devices in the processing chain, such as the multi-screen management system or LED pixel processors, may also lessen the frame delays from those devices. It definitely can be a solution if there is visual tearing or discontinuity between sections of LED wall.

What can generate a genlock signal? Sync pulse generators or master clocks are some of the devices that are used to generate the synchronization signal. A basic test pattern generator that can output analog black can work as well. It just needs to be distributed to all the devices that can take a reference input.

7

PLAYBACK, RECORDS, AND MEDIA SERVERS . . . OH MY

On March 11, 2011, a 9.0 magnitude earthquake struck Japan, centered in the ocean off the coast of Miyagi Prefecture. The resulting tsunami caused hundreds of billions of dollars in damage and more than 19,000 people lost their lives. Japan's industrial infrastructure was dealt a major blow and recovery time was being measured in years.

This was the butterfly effect event that initiated the transition from tape-based playback and recording to digital formats. Although digital recording methods existed prior to this event, their value and proliferation would expand from this point.

As it happened, Sony's main plant for the production of HDCAM-SR and Blu-ray discs was in Tagajo City, Miyagi Prefecture. The factory suffered major damage along with other facilities that produced various types of tape-based media. The impact on Japan's supply chain also forced production to completely shut down.

DOI: 10.4324/9781003247036-7

For the rest of the world, this meant the existing supply of tape media was all there was going to be for a while, possibly ever. Production companies were told to erase and re-use tape stock as the price and scarcity increased. For the live events industry, Betacam and later HD-CAM tape were the main mediums for playing back video content. Occasionally, a presenter would have footage on DVD or Blu-ray that they wanted to include. Usually this was copied to Betacam to make it easier to cue and give countdowns.

There was always a main and backup playback machine in case the tape broke or bound up for some reason. The playback order was linear, so if footage had to be skipped, you were forced to scan at high speed pass the clip to the next. If you were not given backups for your footage, you would dub backup copies and occasionally have to make insert edits to remove footage at the last minute. Today, all playback is digital and has the advantages of being able to assemble playlist, instantly skip to the next piece of content, and duplicate at the speed of a file transfer.

Video Codecs

The bandwidth and amount of data storage required for video clips is based on the codec that is used. Some are optimized for low bandwidth transfer and others are designed for larger uncompressed formats. Usage is based on the quality expected and delivery method of the final product. Codecs are constantly in a state of give and take between bandwidth, quality, and file size.

Compression is an algorithm used by codecs to optimize images and the amount varies based on the sampling type. Intraframe compression describes the process of sampling individual frames of video. Codecs that use this include M-JPEG and can deliver high-quality images but require more bandwidth and are larger files. Interframe compression, such as H.264, samples several frames of video and only tracks the changes in between frames. This results in reducing the amount of data needed to reproduce the video, as well as lower bandwidth, good quality, and small file size.

Image sequences are not a codec, but utilize full frame, high-quality images played back in succession to achieve the ultimate image quality.

This requires huge arrays of fast access storage and high bandwidth computer components. It is the preferred file method for auto shows which demand high detail images of cars and has a unique advantage over the other file types. If a section of frames needs to be altered or changed, only the corrected frames need to be rendered and replaced in the sequence file, whereas the other formats need to have the corrections made and then the entire clip re-rendered.

Deciding what codec to use when recording video is determined by what the end use for that material will be. If the intent is to post the video online using YouTube or Vimeo, then recording in a codec like H.264 would be appropriate as the resulting file would be manageable in size and meet the bandwidth requirement for internet transfer. If the recorded content is meant to be "edit ready," then the format should be an intermediate codec such as Apple Prores or Cineform. Be aware that importing an interframe codec like H.264 into an editing program such as Final Cut Pro or Adobe Premiere, uses more processor power since partial frames of video have to be converted into full frames. Recall that interframe compression only considers the changes in between frames and not the entire frame itself.

Someone is always building a better mouse trap and the same goes for video codecs. As more parameters need to be included (HDR, higher bit depth, higher resolutions), newer codecs appear that help with bitrates through systems and try to reduce storage requirements. One of these is the HAP Codec. Popular as the preferred codec with many of today's higher end media servers, it lowers the data rate with reasonable image quality at its base level. For more critical applications, there is HAP Q which improves the image quality at the cost of larger file size.

Another codec is NotchLC with can be used as an intermediate or playback codec. It adjusts the amount of compression not based on a target bitrate, but by the quality level required. NotchLC is currently only available to Mac users through the Adobe Creative Suite. There is no Final Cut Pro plugin.

Both of these last two codecs take advantage of GPU (graphics processing unit) encoding. A GPU is a separate processor attached to your display card on the motherboard or on a separate graphics card. The processing load of encoding video is handled by the GPU and thus reduces the load on the CPU, the computer's main processor. This can improve render time and throughput for video images.

In case it isn't blatantly obvious by this point, you will need to have above average computer skills in order to successfully be a playback or record operator. Not just in how the different operating systems function but how different processors, ram configurations, and graphics cards affect what the optimum playback codec might be. Be fluent in both PC and Mac and spend some time getting to know the basic editing functions of Adobe Premiere or Apple Final Cut Pro. They use similar techniques of timeline-based editing and they may be useful tools for format conversion.

B.Y.O.H.—Bring Your Own Hardware

Let's talk about the simplest type of media server where you provide the computer and you load a software program onto it that handles the queuing and play out of video clips. QLab (www.qlab.app) is a Mac-based program that has found wide usage throughout the theatrical community. Not only can it play out video clips from a cue list, but it can also trigger audio events and control lighting fixtures as well. Many students learn QLab as part of technical theater programs in high school or college and grow with it into their professional career.

Another Mac only program along the same line is Mitti (www.imimot.com). Developed in Hungary, Mitti is a simple easy to learn tool for video playback. Both Qlab and Mitti, have advanced playback features such as multi-screen output and the ability for warp transformation on the output. In terms of how many outputs they can support is determined by the capabilities of the host computer such as hard drive speed, amount of ram, type of CPU processor, and graphics cards.

For optimum playback these programs recommend using files with a compression format such as H.264 or other file type with low bandwidth and file sizes. Still images can also be incorporated into the cue list and for playout in support of scenic elements and the built-in warp features are handy when using projectors that do not have advanced warping capabilities. A caution when using these programs for external video sources. Processing delays from a capture card, through the software, and to your video output can be on the order of two to three frames. This is before any switching or LED processing downstream.

Renewed Vision (www.renewedvision.com), out of Atlanta, Georgia, has a program called ProPresenter, which is available for Mac and PC, is the primary playout software in the house of worship market. It has been engineered to be easy to use by volunteer staff and simplifies the tasks of titling, inclusion of PowerPoint slides and Bible verses, along with playback of video across multiple outputs. They also developed a program called ProVideoPlayer, which is more geared toward the live events market in general with multi-channel playback and support for large canvases and blended projection. However, it is only available for Mac.

Any one of the aforementioned options is capable of single channel playout at the very least and most offer free versions in order for you to learn the software interface. Past that, performance will vary based on hardware and of course user preferences are a big part of which software package gets used.

Medium Servers

This is not an official category in the industry, but it's the name we'll give to servers that have a software and hardware solution. With their specialized hardware they can handle multiple streams of higher resolutions and have features such as automated backup servers and built-in multi-viewers.

The issue with the "bring your own hardware" solution is that when running into technical issues and having to contact the software manufacturers, they are more than willing to assist with issues regarding the operation of their software. However, they are unwilling or unable to troubleshoot hardware or computer-based issues. They may have you check a graphics card setting or ask what devices you're outputting to, but they will not diagnose your hardware. Also, there are various devices for outputting video signals and audio signals; you may have two or three USB devices attached to your computer to handle this.

Since the computer and the software are bundled together in this category of medium servers, it's to be expected that the software and hardware have been compatibility tested and optimized. Medium servers offer the convenience of an all-in-one hardware and software package with dedicated professional audio and video connectors already on the back of the chassis, but may have a limited feature set which is focused on one to four channels of play out. They often support genlocking, which prevents multiple channels from tearing or becoming asynchronous when butt-edged on display devices such as LED walls.

Some examples in this category are the Coyote from Sonoran Video Systems (www.sonoranvideosystems.com) in Phoenix, Arizona, and the Modulo Pi servers (www.modulo-pi.com), which are based in France. Both systems are sold as a hardware chassis with bundled software package preloaded. The hardware and software were developed in parallel to provide the best performance and offer advanced options like 12G-SDI output.

Super Servers

Again, we're making up categories here, but these servers are in a classification where pixel perfect performance and large canvases are the norm. They come with a price tag to match and are the go-to for any high-profile project. They add advanced features for previsualization, camera-based auto-alignment for warping and dome applications, and extended and virtual reality implementation.

Some mainstream names here are Pandoras Box by Christie Digital (www.christiedigital.com), Green Hippo Hippotizer (www.green-hippo.com), AV Stumpfl Pixera (www.avstumpfl.com), disguise (www.disguise.one), and Dataton Watchout (www.dataton.se). Integrating with external devices and control is a big part of what these servers do. Large sports arena and stadium mapping, building mapping, and theme park attractions all use these mega machines.

Another feature of the super servers is the ability to take external control commands and redundancy. Several models can be controlled by lighting consoles such as the GrandMA via ArtNet commands. Also, an offline server can take over for a unit that has crashed, having been preloaded with all the necessary media. The disguise server's Understudy feature is designed to do exactly that. Although it is not instantaneous, it allows an operator to get a working system back faster than starting from scratch.

8

PROJECTION

What Is Light?

Technically, it is the various wavelengths of the electromagnetic spectrum. Light waves in the spectrum can have periods as tall as a skyscraper to small as an atom. There is only a small region in the middle of the spectrum known as visible light that our human eyes can detect. On either side of the low and high end of visible light, are wavelengths that can be harmful to us. Emotionally, it can be the hues in the sky during a sunset or the coldness of an Artic landscape.

As such, projection can have a scientific element with absolutes and an artistic side based on subjective measurement. Not everyone sees color the same way. Women are said to have more sensitivity to color than men. They see azure and turquoise where men see just blue.

Anatomy wise, both genders share the same instruments for measuring light value and color, the human eye. Light enters the eye through the iris

DOI: 10.4324/9781003247036-8

of the pupil, an opening that expands or contracts based on the amount of light. It then passes through the lens which focuses the image and exposes the light to the retina and photoreceptors called rods and cones. Rods give us a grayscale image and the cones provide the color information. The human eye is the inverse model of a projection system.

Light Source

Projectors can be broken down into several subsystems: light source, optic system, electronic processing, and imaging system. At the heart of all projection systems is the light source. Light output is measured in ANSI (American National Standards Institute) lumens and the process of measurement involves taking the average of nine light meter readings at different points on a certain size screen. For marketing purposes, most manufacturers list their center lumen measurement which is just a value taken from the center of the projection image where the light is usually focused.

In an earlier chapter, we discussed light valve technology and how it led the way to the first light cannons. This is where we will start our conversation about light sources with metal-halide and UHP (ultra-high performance) lamps. The difference being that UHP only contain mercury versus it combined with a metal. As the source for these early light cannons, UHP mercury lamps are compact in size, easy to install and have low power consumption. Brightness is increased by adding multiple UHP lamps together. However, these lamps have an efficiency ceiling which means that after a few hundred hours of operation, the light quality and color accuracy start to degrade as they approach their maximum lamp life of about 1000 to 2500 hours. Then there's the issue of disposing these lamps since they contain mercury under extreme pressure. Although UHP lamps are still used today in lower lumen projectors for small meeting and boardrooms, we are seeing a transition to LED illumination or laser phosphor in these applications.

Xenon lamps have been used in projectors to attain lumen values of 15,000 and higher. Xenon requires special handling and commonly runs on 208 volts minimum. This means in the United States and other countries with 120-volt single phase, it requires two phases of power to operate the projector. Failure of a xenon bulb can be very explosive, so the lamp housings are built to contain debris and in some cases the lamp must be transported separately from the projector. Although very bright, xenon light can be very uneven due to the nature of arclight. As the position of the arc moves between the electrodes in a xenon lamp, it changes the light output and centering within the reflector, a mirrored dome that reflects the light out of the lamp housing. It is therefore necessary to perform what is called a Z-axis alignment about every 100 hours to optimize the arclight's position in relation to the reflector. It is surprising how many projectors with reduced brightness levels can be increased with this simple procedure.

Since only one xenon lamp can be present in a projector, usage of two projectors double stacked where one is a backup in case of failure, is common. With UHP lamps, most higher brightness units have more than one lamp, so a failure of a lamp does not result in total loss of image, just a dimmer one.

Recently, projector manufacturers have been switching to either a full laser or laser phosphor-based illumination systems. This is highly efficient, has a much lower failure rate, and has a built-in backup system much like UHP lamps. The trend to switch to laser illuminated projectors began with digital cinema. The need to produce images at higher brightness levels and frame rates, particularly for 3D movies, was the driving force behind this change. The largest projectors made for digital cinema replaced the lamp completely with a three-laser system for illumination. Keep in mind that lasers are being used to generate white light and not exit the projector like in a laser light show. The method for converting RGB lasers into white light is the trade secret for each manufacturer. It is this laser-created light that travels down through the usual optical and imaging paths to produce an image in the traditional sense.

In order to cool these laser units, additional chiller units had to be incorporated that are external to the projector and require extra power. This is not much of an issue in a fixed installation, such as a movie theater, but for rental staging and live applications, a true RGB laser system with chillers is not practical and would probably not hold up to the abuse. For these applications, laser phosphor illumination is able to produce similar lumen output while still maintaining a reasonable form factor. In some cases, being smaller than their xenon lamp equivalent.

Unlike the RGB lasers used in digital cinema, laser phosphor uses a lower power blue diode laser, similar to the ones in Blu ray players. The diodes are grouped into blocks which can have anywhere from four to eight diodes. Failure of a single diode does not cause all the other diodes to fail. In most cases, the group containing that diode will go out, but the remaining groups will stay on. This is a built-in redundancy system which in the worst case dims the image, but there is still an image on screen to be seen. Double stacking projectors for the sake of a live backup is somewhat unnecessary with this technology.

But how is white light achieved from blue laser diodes? That is where a phosphor wheel comes in. Energy from the blue diodes excites the phosphorus coating on the wheel, generating green and yellow light. Through a series of dichroic mirrors, red and green are extracted from the phosphor light and blue takes a separate path to meet with the other color components to become white light. How the light is processed from this to point depends on the number of imagers. For single DLP chip projectors, the light hits a color wheel which spins at thousands of RPMs and has segmented sections for red, green, blue, clear, and yellow. Through synchronization with the single DLP chip, the appropriate color is matched with its portion of the image.

For projectors with three DLP chips, the combined laser phosphor light enters what's called the light pipe and is reflected into a prism system which distributes the appropriate red, green, or blue colors to their matching DLP chips.

There is less heat dispersion with the laser phosphor system and many are liquid cooled. This means smaller fans or fans with lower speeds can be used to cool the projector. The result is a quieter unit. Unlike xenon projectors, laser phosphor projectors can be pointed in any angle because there is no anode or cathode. The orientation of the anode and cathode on a xenon lamp is important and if it is positioned incorrectly, you will experience flickering or reduced light output. Laser phosphor also produces a more even light output from edge to edge. To date, many projectors have gone years without the need to service the laser diode modules.

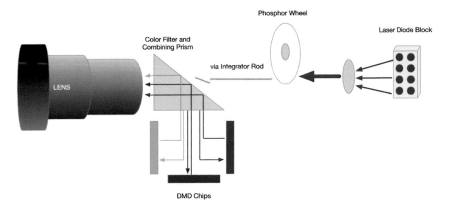

Figure 8.1 Illustration of three-chip DLP system using laser phosphor light source.

It is important to note that any type of laser in a consumer or professional product is regulated in the United States by the Food and Drug Administration (FDA). As such, laser phosphor projectors are classified as laser show devices and have a class and risk number assigned to them based on their laser output rating. Most projectors fall into the Class 3 laser product category. What this means is that manufacturers of laser phosphor projectors must apply for what is called a variance in order to sell their products in the U.S. This variance ensures that the laser power is contained within the unit and no direct contact with the lasers is possible. Likewise, any distributors or resellers of laser phosphor projectors must also file a variance with the FDA. Operators of laser phosphor projectors are only required to take a safety program and be able to show proof of completion.

Through our description of the technology, you can see that no lasers are emitting from the projector, it is converted into white light that leaves the projector the same way it has since the beginning of projection technology. Furthermore, these requirements of registration and variances are not required anywhere else in the world outside the U.S.

Optics System

In a multi-UHP lamp or multi-LED module projector, the first optical device is a combiner. This takes the aggregate light from all the sources and directs it towards the imaging system. The light passes through dichroic mirrors on its way to split into the various colors. On the other side of the imaging system, light is recombined and exits the projector via the lens.

With a xenon system, the first light filtering happens at the light housing where coated glass filters out the UV rays from the light. The next optical component is what is referred to as the "cold mirror," another piece of coated glass that filters out the infrared wavelengths from the light and dissipates their heat via a heat sink. What is left is the visual light spectrum which is concentrated on the end of the integrator rod. The rod is not round but actually rectangular and in the same aspect ratio as the DLP imaging chips, called DMDs (digital micromirror device). Like a huge fiber optic pipe, light travels down the integrator rod and bounces off a fold mirror into a prism block that divides the light into red, green, and blue.

The colored light hits its associated DMD and bounces back into the prism where the beams recombine and again exit the lens. Over time, xenon systems can require adjustments to members of the optical system. The cold mirror may need realigning, the integrator rod may need to be turned slightly, or the fold mirror which sends light into the prism block may need adjusting.

As laser phosphor is merely a replacement of the light source, the optic system in these projectors will look very similar to either the UHP or xenon model, with the exception of those parts that remove harmful light rays.

Electronic Processing

Texas Instruments, the maker of DMD chips for DLP projection, controls the specifications for the resolution and maximum lumen output for a given series of chips. All projection manufacturers using DLP technology receive the same chipset from Texas Instruments. It is up to the individual manufacturers to develop the electronic processing systems that format incoming signals to the DMD.

In terms of digital cinema, DCI specifications require that there is no image manipulation between the source server and the formatter for the DMD. What this translates to is there can be no scaling, pixel manipulation, or warping of the source material. But these projectors are application specific and cannot handle any external miscellaneous sources without the use of an electronic interface.

For projectors used in the live events and staging industries, the projectors come with a built-in set of electronics that handles format conversion, scaling, warping and even color correction. Connectors are generally found on the side of the projector on what is referred to as the e-box or electronics module. Depending on the capabilities of the projector and resolution of the DLP chips, there may be an assortment of connectors available for input sources. The most common are HDMI, DisplayPort, and SDI.

Controls for the projector are accessed via a menu system and settings and changes to the configuration of a projector can be saved to internal memory or exported to a USB stick. Direct access is always given through a keypad on the side of the projector or through a wireless infrared remote. It is not uncommon in configurations with multiple projectors to have all the units networked on a VLAN. More about networking will be covered in a later chapter, but what this simply means is that all the projectors can be controlled from one computer, either wired or via Wi-Fi connection.

Use of a Wi-Fi network is common in medium to large projector setups where it might be necessary to walk around the venue and adjust projectors hung in the air. However, this Wi-Fi network should be disabled and converted to a hard wire once actual show days start. The presence of thousands of Wi-Fi connected devices or devices looking to connect to your projection network, may cause serious issues in terms of wireless traffic. It should also be mentioned that automatic IP addressing is not a good practice when setting up projector networks. The use of static IP addresses ensures that your projectors can be identified by their location every time the software tries to connect with them.

Scaling chips within the electronics module are used to resize input sources. In the case of an HD 720p signal coming into a projector that has a 1920 × 1200 native resolution DMD chip, the scaler would be used to scale the 1280 × 720 input signal to fill the chip, while maintaining the input's aspect ratio. In doing so, it tries to utilize the maximum light output of the projector by using as much of the DMD surface as possible. There are some use cases where mapping a signal natively versus resizing can be useful. Using the same example of a 720p signal as an input, let's say for some reason you received the wrong lenses, and your projected image is overshooting your screens by a considerable amount. By setting the projector to map the signal natively, a 1280 × 720 resolution image is mapped in the center of the DMD chip. This is of course a smaller image and less bright, but it is now closer to the screen size you were trying to attain. Also, natively mapped resolutions are pixel perfect and will not have the added softness created by scaling them up in size.

A more practical application for selecting native mapping is to reduce processing latency that may be caused by routing signals through the scaler. If an input signal matches the resolution of the projectors DMD chip, then it can be mapped natively. It is a good time to note that any processing changes you make on one projector in your system need to be made on all the projectors in your system. For example, when blending two projectors together for a widescreen, if one projector has native turned on and the other does not, there will be a processing delay and images in the center of the screen will tear or disconnect.

In some cases, the electronics box is powered up when the main power is connected, and some functions and settings can be completed without fully turning on the projector. Examples of these are setting the IP address or initiating a factory reset. Of course, the projector needs to have some sort of display in order to do these at the projector, otherwise they can be done via built-in web pages or remote software. Again, if your projectors are on a network, you can often reach a web page control menu by simply typing in the IP address of the projector into a web browser.

Imaging System

There are two main imaging systems that are used in today's modern projectors. The first is LCD or liquid crystal display. Developed in the late 1960s, liquid crystal displays would find use in digital watches, calculators, and early computer displays. It wasn't until the late 1980s that LCDs were first used in projectors.

TFT or thin film transistors are the technology behind LCD. Single pixels can be turned on or off, or some value in between, to allow light to pass through the LCD panel creating a light valve. Typically, one panel is used to process each of the primary colors red, green, and blue. This is transient light processing as the light is passing through the imager and out through the lens. As opposed to DLP technology where the light reflects off the imager.

Today the main manufacturers of LCD panels for projector use are Epson and Sony. Several years ago, Epson introduced a series of projectors based on laser phosphor illumination. These units have lumen output of 12,000 and above and have image quality rivaling a comparable DLP. To date, there is not a native 4K LCD imager. However, since Epson owns the technology and controls the manufacturing process, it is possible since they are not handcuffed to a chip supplier like their DLP competitors.

The prevalence of LCD projectors in the live event space may be based on pre-formed opinions of earlier technology. Historically, LCD has had a limited lifespan due to the discoloring of the polarizers, in particular the one for the blue panel. This causes a yellowing effect and can only be fixed by replacing the panel. This happens within three to five years of projector operation. Epson has claimed to have corrected the issue by using different materials for the polarizers, but the stigma of LCD as a technology for smaller applications still exists.

Digital Micromirror Device

In 1987, Dr. Larry Hornbeck of Texas Instruments created the digital micromirror device, or DMD. Both an opto-electrical and electro-mechanical device, the technology would bring about a paradigm shift in how we view movies and project images. Dr. Hornbeck's creation would receive due recognition in April 2015 with a Technical Oscar from the Academy of Motion Picture Arts and Sciences.

As the imaging system in a DLP projector, the technology is a reflective system since the light entering the DMD bounces off the microscopic mirrors and into a combining prism before exiting through the lens. On a microscopic level, there is a micromirror for every pixel in the image. This mirror pivots to either a positive or negative direction depending on whether that pixel is on or off for a given color. Try to imagine a large solar panel farm and that each solar panel represents a pixel. With the sun as your light source, if it hits the solar panel and that pixel is active in the image, then the panel reflects light towards the lens. if that pixel is off, then the panel reflects light away from the lens. Light that is not used as part of the image gets directed to what is called a "light dump." The light dump has a heat sink to draw away the heat created by the light. With this principle in mind, it is important to note that when a projector is outputting black, what that actually means is that all the light is being directed toward the light dump.

Projectors that are left outputting black for an extended period of time can have overheating issues. It is better to leave a white signal going to the projector and use the shutter or project white without the use of the shutter if the projector needs to be left on but not actively in use.

However, this only applies to projectors that use a xenon-based lamp source. The newer laser phosphor projectors when used with DLP technology do not have this issue necessarily since when the shutter is engaged, the laser diode banks are turned off. They instantly turned back on when the shutter is disengaged. This instant on/off of the laser diode banks is another advantage that the laser phosphor technology has over traditional lamp-based projectors. Images appear on screen faster and there is nearly no cooldown time required when powering the unit off.

DMDs are also being produced in native 4K resolutions and projectors using these native chips have achieved lumen ratings of up to 60,000 with RGB laser light sources.

4K Sleight of Hand

Before imagers were available in native 4K resolutions, manufacturers had a little trick up their sleeves that they would use to create 4K images from imagers that were of less resolution. The technique is commonly called pixel shifting or wobulation.

Typically, the imager size used was WUXGA (2560 × 1600) resolution. The exact details of how each manufacturer did their pixel shifting was their own trade secret. But the theory of it goes something like this. Somewhere in the optical path after the combining prism and the entrance to the lens sits the pixel shift device. When the projector is being asked to output a 4K image, the pixel shift device goes into action and not only creates images from the native pixel positions of the DMD but also directs light one pixel up and one pixel to the left, creating virtual pixels on screen. It scans through these positions about 100 or so times a second. Due to your eyes' persistence of vision, it sees the pixels being created but does not detect the motion of the pixel shift device. So, in essence, every 100th of a second a third of the pixels that make up the image are physically not there.

Though it may sound like cheating, this technique produces quite excellent results with regards to 4K imagery. Some have reported issues with double stacking or warping overlapped projectors due to an illusion that alignment grids appeared to be shifting. This is understandable since there is no way to synchronize the pixel shift creation between two projectors.

To Stack or Not to Stack

The longstanding technique of double stacking projectors is done for a couple of reasons. One is in case of lamp failure. With highly volatile xenon lamps, lamp swaps cannot be done immediately after a lamp is blown so having a second unit aligned to the same image that can take over is one reason for double stacking. In these cases, sometimes the second projector is aligned but shuttered, this is called a "paused standby." If the second projector is on, then it is a "hot" or redundant projector.

The other reason for double stacking projectors is to increase brightness. Some situations may require the projection to overcome some ambient lighting or heavy stage lighting effects. It is always a good idea to check with the technical director or the account executive for your event to see how the projectors were quoted and are going to be billed to the client. If they are only paying for a non-active backup, then you are doing a disservice to the projector owner by running that projector for more hours than are necessary. In the case of an edge-blended projection screen, you will have to do the work anyway to align the backup and the main projectors whether both are on or not.

If your physical geometry in placement is correct, you should be able to align the bottom projector to the screen without any electronic warping. The projector on top should be able to align very closely with the bottom projector and may only require a few clicks of corner correction. When teaching students how to double stack projectors, I always remind them of the 80% rule: you can never get projection to be 100%, but 80% is good enough for rehearsals. The spirit of this is meant to not waste your time obsessing to get one pair of projectors 100% dialed in. Get it to 80% and then move on to the next pair. As you find time, go back through and try to get that last 15%.

Another trick that comes in handy when double stacking or any kind of overlap projection is the use of RGB cutoffs. Mostly available in projectors that have three imagers, one for each of the primary colors, it allows the user to turn off one or more of the colors being projected. Therefore, if using a test grid and one projector has only the red color enabled and the projector you're trying to align it with only has the green color enabled, then when the two grids overlap the line color will turn yellow. Likewise, if one projector is red only and the other projector is blue only, then their overlapped lines will be magenta.

Widescreen Edge-Blending

In the late 1970s and into the early 1980s, widescreen presentations using multiple stacks of 35 mm slide projectors were not uncommon for large corporate events. These included annual stockholder meetings, model reveals for car manufacturers, and presentations purely for entertainment purposes. The show itself would take months of programming, and on site projectors were aligned using standardized registration slides. These projectionists only had the ability to manipulate the physical position and geometry of the projectors; there was no electronic adjustment.

Years went by, and due to the cost and time needed to produce these kinds of presentations, they were eventually replaced by multiple single screens using video projectors connected directly to computer outputs. Gone were the days of having to rescan slides in order to make corrections and then trying to find an overnight photo lab that could process them. Now, graphics could be updated on the computer itself up to the moment the presenter walked on stage. Multiple screens that switched between I-mag (image magnification) and PowerPoint presentations was the norm for several years and systems were developed to make multiple screen management easier. This coupled with the advancements in projection, in particular DLP, and the need for producers to come up with something new for their clients' events, led the industry to revisit the multi-image widescreen format.

In the beginning, projectors were outfitted with physical devices to create the feathered edge since at that time they did not have software internally to do this and the multi-screen systems and processors were lacking this feature as well. The other half of the blending puzzle was coming up with a way to create data doubling, a method of sending an image to a particular projector that contained portions of the overlapped content. Both of these problems would be solved by the high-resolution processors of Folsom Research (Barco), Analog Way, and Vista Systems.

One of the biggest questions always asked was: "How big does the overlap need to be?"

The overlap between projectors is expressed in pixels and there seems to be some confusion about what is considered a good percentage of overlap. In the early days of widescreen projection, the percentage depended on what projectors you were using. DLP projectors which had a more uniform edge to edge color, could get away with smaller percentage overlaps. Whereas UHP lamp projectors with uneven color from left to right, would require larger overlap areas to hide their imperfections. In fact, it was highly ill-advised to use UHP lamp projectors for wide screen blending applications for this reason.

But getting back to calculating the amount of overlap pixels, the answer is found using math, in particular pixel math. Let's say you are trying to figure out the overlap and how many projectors are going to be needed on a screen that is 10 feet tall by 30 feet wide. The projectors you are using will be outputting a 1920 × 1080 resolution image.

Step 1: Calculate the size of one projected image.
Since the resolution of the projector is 1920 × 1080, the aspect ratio of the image is 16:9.16:9 as a decimal is 1.78 (16 divided by 9). The height of the screen is 10 feet.

10 × 1.78 = 17.8 The size of one image is 10 feet tall by 17.8 feet wide.

Step 2: Use the size of one image to figure out pixels per foot.
Since the height of the image does not change over the distance of the screen, 1080 pixels is the height all the way across. Horizontally, the distance increases. Knowing that one image in 17.8 feet wide, we divide 1920 (the horizontal pixels in one image) by 17.8.

$$1920 / 17.8 = 108 \text{ pixels per foot.}$$

The calculation actually returned 107.865, so we rounded up to 108 since you cannot have partial pixels.

Step 3: How many pixels exist across the horizontal axis of the screen.
Taking our 108 pixels per foot and multiplying by the 30 feet of horizontal distance:

$$108 \times 30 = 3240. \text{ So there are a total of 3240 pixels horizontally.}$$

Step 4: How many projectors does it take?
One projector horizontally is 1920 pixels. Therefore, two is 3840 (2 × 1920). 3840 is bigger than 3240, so the answer here is two projectors.

Step 5: What is the overlap pixel amount?
Taking the two projectors' total horizontal pixels, 3840, and subtract the screen's horizontal pixel count, 3240.

$$3840 - 3240 = 600. \text{ Six hundred pixels is the overlap pixel amount.}$$

That works out to be about a 30% overlap between the two projectors which is more than enough. But what if we change the vertical resolution of the projector to 1200 pixels? You can run this back through the calculations above and you'll get a number much lower than 600 pixels. However, if using DLP projectors, it is still a very descent overlap amount.

In our example, the screen size of 10 feet × 30 feet is a 3:1 aspect ratio. The overlap pixel amount will be the same for all 3:1 aspect ratio screens, so a 20 foot × 60 foot screen would also have 600 pixel overlap when each projector is outputting 1920 × 1080.

Geometry

Nothing saves more time when setting up projectors then correct geometric placement. In a projectionist's tool bag, there are a few key crucial items: two 100-foot tape measures, a laser distance meter, a crescent wrench (spanner), and screwdrivers, binoculars, and a small bubble level or laser level.

The first procedure would be to measure and find the center of your screen. Next triangulate the projectors position in relation to the corners of the screen. Your projector's distance from the screen will be determined using a lens calculator which we will go over in the next section. Using the tape measures, have two people hold the end of the tape measure at the bottom corners of the screen. Then have the tape measures cross at the front of the projector. Look at the numbers on each of the tape measures and if they are the same, or at least within a few inches of each other, then you are centered with the screen. If one tape measure is longer than the other, then shift your projector in the direction of the longer tape measure Until you are within that one- to three-inch tolerance. This does not have to be exact because today's projectors have enough lens shift to compensate for not being in the perfect position. However, if you can keep the image centered in the lens without having to do much shift, you avoid aberrations or distortions that may happen as the image shifts towards an edge of the lens.

If by yourself, you can use a laser distance meter in the same way. Just be sure to have the laser distance meter at the projector pointing away towards the screen. Never have the laser pointing at the projector! Laser light entering the lens will cause instantaneous and permanent damage to the DMD imager. It doesn't matter if the projector is on or off and unfortunately once you discovered you have the damage, there's nothing you can do on site to correct it.

This is especially true for those live events that involve laser shows. Much care needs to be taken by the laser operators to not have their beams intersect with any of the projectors. A good standard operating procedure would be to have projectors shuttered while lasers are tested and aimed and only unshuttered after viewing the laser presentation to ensure that no lasers are aimed at projector lenses.

You will find that by taking the time to properly set your physical geometry, your alignment of the main projector and subsequent alignment of the second projector in the stack will be much easier and not require a large amount of electronic warping.

A note with regard to warping on curved surfaces. Be sure that you have enough range with the lens to slightly overshoot the curved objects. You can always warp in from the edges, but you can't put pixels out past the edge of the imager.

Lenses

Today's projectors have a wide range of lenses available to them, including those that have special purposes. The lens most commonly used is a zoom lens. It has an adjustable focal range and is the best option when dealing with widescreen projection or warping on surfaces. You never want to be at one extreme or the other of the lens. Plan your throw distance to be somewhere in the middle range of the lens, this way you have the ability to grow or shrink your image for alignment.

When using short throw lenses, be aware that they have a fixed focal length and need to be in a precise position, commonly center screen. This means when being asked to double stack projectors with short throw lenses, one projector will be in the optimum position while the second projector will be far from optimal and will require a large amount of electronic warping to overlay with the other projector.

Another lens you may encounter is the ultra-short throw lens, also nicknamed a "snorkel" lens. These lenses incorporate a mirror which reduces the amount of throw distance required to less than a one-to-one ratio. One such lens, used by Panasonic projectors, allows them to sit a few feet in front of the screen surface and still fill screen sizes up to 16-feet wide. The downside is that in this configuration the lens is shooting back over the body of the projector. The narrow light cone does not allow for double stacking of projectors. The only method of having two projectors on the same screen with these particular lenses, is to have one projector shoot from the bottom of the screen and the other from the top of the screen. It is often impractical and unnecessary to do this and most people find the single projector is bright enough.

Other manufacturers use a lens which shoots sideways using a 90-degree angle. Projectors can now be double stacked and have some range of lens shift available to them.

Lens calculations are made using three variables: focal length, distance, and screen width. If you know any two, you can figure for the third. Here is an example. You know your screen is 16 feet wide and you're using a zoom lens with a range of 1.3 to 2.0.

Projection distance = Screen width × Zoom factor
$$X = 16 \times 1.3.$$
$$X = 20.8. \quad \text{Minimum projection distance required is 20.8 feet.}$$
$$X = 16 \times 2.0.$$
$$X = 32. \quad \text{Maximum projection distance is 32 feet.}$$

If you had a 20-foot wide screen and were shooting it from 100 feet away, what size lens would you need? If you said something including a zoom factor of five, you'd be correct (projection distance / screen width). And who said you'd never use that high school algebra ever again.

All the major manufacturers of projectors have lens calculators either online or a separate application that allow you to enter two of the three values to get the third. These calculators are much more accurate than the simple math we just completed because they include variables and offsets that are particular to their lenses and may increase or decrease the projection distance of that lens. They can also take into consideration whether you are projecting a 16:9 image versus a 16:10.

Projector Safety

The projectors used today in live events are complicated machines costing tens of thousands of dollars. But the cost of injury or death by mishandling or improperly securing these units is immeasurable. Be sure to follow the manufacturers specifications in terms of the projectors known rigging points, projector weight, and operating environment tolerances.

The use of rigging cages that are designed for a specific model and tested to hold the weight is highly recommended over foregoing the cages for convenience or to save weight. This often causes conditions to manifest that reduce your ability to make precise adjustments and can be a source of frustration, especially when hanging in the air and having to access from a scissor lift.

Take into consideration when having to lift projectors up into position on scaffolding. A four person lift of a projector to the first position may only be a six-foot lift, but the next projector will have to be an overhead press to seven feet. Consider using a forklift or rigging a single motor point backstage and having your scaffold on wheels, so that you can lift projectors easily to the necessary heights.

Power is another safety concern. As mentioned before, some of these units require 208 volts or higher to operate. Here in the United States, and anywhere with single phase 120-volt power, 208 volts is created by combining two 120-volt lines from separate circuits or phases plus a ground. If you are unsure about how to create 208 volts from single phase 120-volt systems, consult with an electrician or a more experienced lead on your show crew.

In most cases, projection power is run over long distances. This requires the use of large gauge cables in order to reduce voltage loss. A 20-amp, 218-volt power source can drop to 203 volts at the end of a 200-foot cable run using ten AWG cables; American Wire Gauge standardizes electrical cable sizes. Cables are called out by their gauge and number of conductors. Examples are 12 × 3, 12-gauge times three conductors, and 10 × 3, 10-gauge times three conductors. The cable gauge required is tied to the amperage draw of the projector. Twelve-gauge cable is rated for 20 amps while ten gauge is rated for 30 amps. All projectors are required to list their amp requirements near the power inlet or on the manufacturer's plate.

A handy formula to know if you are going to work with electricity is Ohm's Law.

Amps = watts / volts.

Yes, algebra again. There are many iterations of ohms law, but this is the simplest formula to memorize and again if you know two of the variables, you can figure for the third. So how many amps does a 500-watt lighting fixture use?

Amps = 500 / 120.
Amps = 4.17.

In this equation, volts were 120 because we are the United States. Each 500-watt fixture has an amp draw of 4.17 amps. That means on a typical 20-amp circuit, you can plug in four of these lighting instruments. Plugging in five units would be just over the rating of the circuit.

A device you might see in the field is called a buck booster. This is usually a grey electrical box that is very heavy and has an inlet and an outlet connector with a 20-amp rated plug. Inside the gray box is copper wiring wound to create an induction coil that boosts the voltage of one leg of your power feed. One phase enters and exits the box at 120 volts and the other phase enters at 120 volts and exits at about 140 volts. The result is the ability to boost a 190-volt feed to 208 volts after running power over a distance of 200 feet.

Buck boosters should always be used at the projector end of a power cable, never at the power source end. Also, you should measure the power at your projector with a multimeter before installing a buck booster. They should not be installed without first making a voltage measurement. If the voltage measurement at your projector is with between 208 and 220 volts without the buck booster, this is adequate and the buck boosters should not be used. If less than 208 volts, then the buck booster can be installed safely. Buck boosters should also not be used on any power runs that are less than 100 feet. Most projector power supplies can handle voltages between 100 and 240 volts with the optimum being around 220 volts. Overvoltage to a projector can be just as damaging as undervoltage and both conditions can cause cables and connectors to overheat, leading to possible fires.

Lastly, is laser safety protection. As mentioned earlier, laser phosphor projectors are regulated by the Food and Drug Administration (FDA) in the United States. They require that operators take a safety course in the safe installation and operation of Class 3 rated laser projection devices.

This requirement does not exist anywhere else in the world where laser phosphor projectors are classified the same as any other lamp-based projector. Does the FDA lack logic and critical thinking skills to realize that lasers are not emitted from laser phosphor projectors? The precedent for this goes back several decades when the FDA was put in charge of all lasers, whether used for medical or entertainment purposes. This includes the use of lasers in CD and Blu-ray players. The FDA's concern with laser phosphor projectors is that higher lumen outputs are being achieved in smaller unit sizes and that human's natural protection system, the ability to blink, may not have the reaction time to prevent damage when a person accidentally looks into the projection cone of these higher lumen devices. This is especially concerning when video projectors are used to present material to young children.

For live events, it is important to set up safety zones so that crew, participants, and the audience are prevented from accidentally being exposed to the light cone within dangerous distances. Backstage, projectors should be at a minimum height of 2.5 meters or about 8 feet off the ground if the projection cone area is going to be used as a walking path. Otherwise, the projection cone area should be marked off so that no transient traffic is allowed.

Likewise, the first three feet of the projection cone starting at the lens, should be avoided. Should it be necessary to work in this area, the projectors should be shuttered or turned off until this three-foot safety zone is clear again. This last requirement probably is more appropriate for festivals and concerts, and that is to have a spotter who can remotely shutter projectors in the event that someone breaches the safety zone areas.

Screens

What is a projector without something to project onto? In most cases, they are projecting onto screen material and the parameters and specifications for screen surfaces and materials varies widely.

The most common types of screen fabrics are what are called front and rear surface. Front surface material is typically white with a reflective coating and are available in roll up and fast-fold configurations. Rear surfaces are typically a gray, vinyl-like material that is translucent and allows light to pass through and mostly comes in only the fast-fold configuration. Fast-fold describes a system using a frame assembled for multiple parts onto which the projection surface is usually attached using a series of snaps. Larger frame systems may use a lace and grommet system for attaching the screen fabric. When ground supporting these fast-fold frames, leg kits are added which attach a T-shaped leg base that has multiple positions for height adjustment. Large format widescreens are flown from pipe or truss due to their weight and lack of center supports.

An important tip. Screen fabric tends to be more pliable and stretch easier when at room temperature or above. if you're installing a screen in the middle of winter or it traveled overnight in the back of a very cold truck, I would suggest unpacking the screen fabric and letting it heat up for a little bit before trying to attach it to the frame.

The reflectivity of screen material is measured by what is called screen gain. Screen material with a gain of one means that there is a one-to-one ratio between the light from a projector and what is reflected off the material. A screen gain of 1.5 means 50% more of the light is reflected off the surface and a screen gain of .8 means that only 80% of the projectors light is reflected. This is a case where having a higher number may not be better. Higher gain screens tend to have narrower viewing angles, so their use where much of the audience is offset to the screen it is not advisable. Screens with a gain lower than one have wider viewing angles and are usually assigned to rear surface material.

The gold standard for rearview ultrawide surfaces comes from Stewart Filmscreen. Their Aeroview 70 series has a gain of .7 and a 120-degree viewing angle. Stewart screens are created by pouring material into form equal to the size of the finished screen. It is cured and then rolled up. Folding these screens would create permanent creases and make the surface unusable. Any company renting these screens will send a screen technician to make sure it is handled and packed away properly.

Do you remember earlier when we mentioned that widescreen projection started to make a comeback years after the end of slide projectors? Well, many of the companies that still had widescreen surfaces from those slide projection days, tried to reuse those surfaces when doing video projection. The problem was that those surfaces were designed for much lower lumen projectors and the higher lumen video projectors were accentuating all the flaws and aging in the material.

Today, companies such as Da-Lite and Draper produce wide format screens that are designed for today's higher lumen output projectors but have a lower price point and higher availability then Stewart Filmscreen.

9

LED

The beginnings of what we know today as LED display technology began in the late 1980s. Sports stadiums and arenas were installing large video displays and scoreboards that could show up close action. There were three main competitors in this area: Sony with the Jumbotron, Mitsubishi with Diamond Vision, and Panasonic with the Astro Vision system.

Although all three continue to have success manufacturing LED display products, it was Sony's Jumbotron, introduced at the 1985 World's Fair, that cornered the sports market and was the first LED video display in Times Square in New York. It's modular panel system allowed for maintenance and scalable size from large stadium installations to smaller mobile LED trucks.

Eventually, Sony would introduce Daktronics to the sports venue market and eventually hand off that business completely to them. The term "Jumbotron" became ubiquitous for any large LED videowall, and it would be more than a decade before the technology of LED displays would evolve.

DOI: 10.4324/9781003247036-9

The Irish band U2 has always been known for pushing the visual and technological envelope with the stage designs used on their world tours such as "PopMart," "Zoo TV," and "360.". Their stage designers, Willie Williams, and the late Mark Fisher, came up with concepts for how LED could be used to display large scale video. They needed a manufacturing partner and they looked across the English Channel to a Belgian company called Barco. Barco had been producing modular video panels for use in live events and did all their manufacturing in-country. This collaboration would create some of the most unique LED visual display hardware ever created and would inspire other companies, such as Element Labs, to look at LED in ways other than just large television sets.

In the early 2000s, Barco had few rivals as they continued to innovate in the LED display market with semi-transparent LED screens, 4 mm-pitch modular tiles that attached to lightweight carbon fiber frames, cableless interconnections, and 12-bit processors. But the price tag for this innovation meant only a few companies with deep pockets could maintain an inventory. When pricing out a LED solution versus blended widescreen projection, projection was always cheaper and still had good bang for the buck.

Then the Great Recession of 2008 hit. LED sales for Barco came to a grinding halt. Those companies that were still looking for LED products turned to Chinese manufacturers that could produce low-cost versions of other manufacturers' products or help create custom products. Production and rental companies turned into LED manufacturers.

The business model was something like this. You had a tour that was going to out for six months. You had the exact amount of LED products manufactured in China to cover the tour and you charged your client just enough to cover the costs. At the end of the tour, you would scrap the LED.

Some owners decided to keep the LED and try to re-rent it on corporate and other events. What they ran into were issues trying to repair or cross-rent their panels since their product was unique and did not batch match with anyone else. For the Chinese manufacturers, this brought about a boom that has made them the de facto source for LED displays worldwide.

Component Breakdown

LED panels begin with their namesake, LEDs. There are several types of LED arrangements that have been used. Discrete LED means that there is a separate diode for each color: red, green, and blue. Because of the physical size of the diodes, the clusters cannot be placed that close together. In fact, discrete LED was used primarily for outdoor rated LED screens with a pixel pitch of 10–12 mm. Pixel pitch refers to the physical distance between LED diode clusters. Discrete LEDs were also able to produce high brightness levels for outdoor displays to compete with daylight. The brightness of LED is measured in nits (a measurement of direct light) versus lumens (a measurement of reflected light). Outdoor panels can be rated up to 10,000 nits, while indoor panels are around 1200 nits.

SMD, meaning surface mount diodes, have all three colors contained in a "package." This is the majority of the LED in use today. The space in between colors is filled with either white or black resin. Black resin can help to increase contrast levels or give the face of the LED wall a darker look. Pixel pitch with SMD has gotten down to 2.6 mm. In order to get closer to the 1 mm pitch, we need to consider COB, or chip on board, LEDs.

Chip on board is where the color LEDs are mounted directly on a circuit board without any package like SMD, allowing for tight spacing. Voltages to the individual LEDs can be better controlled than in SMD and this leads to a reduction in power consumption and lower operating temperatures. A downside is that the modules of COB LEDs are covered in a protective coating. Should a COB LED go out, the module must be removed and discarded. This is more economical than trying to replace the damaged LED.

Not All LEDs Are the Same

Like with many other things, LEDs are available in varying levels of quality. The highest quality LEDs come from Nichia in Japan. Companies that source from Nichia are looking for high brightness and excellent color calibration across the batch. This of course makes them the most expensive selection and products with these LEDs are priced accordingly.

Second in line, and probably the most widely used, is Cree out of China. This is a high-quality LED with consistency across batches and good working lifespan. There are many others after these two main brands, but they all clump together into a generic pool of manufacturers.

Individual LED modules are created in what are called "batches," and millions of LEDs can make up a batch. When a tile manufacturer orders the LED for their product, they already know how many tiles are going to be created in that production run and how many LEDs they need. Once that product run is complete, all of it may be sold to a single customer or divided among multiple customers. If a year later any of those customers wants to purchase more of the same product, chances are the LEDs in that new product are a different batch from the ones they own. This makes color calibration between the two series of product very difficult. It usually involves individually color calibrating each panel using specialized cameras and software in a dark room.

This is hardly the case on show sites which makes field calibration nearly impossible.

If you know you are using two different batches of tiles ahead of time, it would be best practice to mark the cases for each batch accordingly. On show site, build one section of the wall using one batch and complete it with the other batch. otherwise, if you accidentally mix batches together in the same section, your only recourse is to try and calibrate (color match) the tiles by eye, a long and laborious process that usually doesn't have the best results. An even better solution would have been to locate the other owners of the same batch and cross rent any needed extra tiles.

Sending and Receiving Cards

Sending and receiving cards handle the communication between the pixel map processor and the individual modules that make up an LED tile. The term "sending card" is a bit of a holdover from when a card in a PC slot of a computer sent data to tiles. Nowadays, this is handled by stand-alone dedicated hardware with onboard firmware. The walls can be configured directly on the front panel of the device. More often, though, a computer is still connected to the pixel map processor to define configurations, run calibrations, or do any firmware updates required for the processor or the tiles.

The receiving card is a small PC board located on the back of the lead tile. It is usually housed in what is termed as either the "hub" or "core." These hubs are the passthrough points for power and control signals and can be removed for easy servicing or troubleshooting. A main component to the receiving card is the driver IC, the chip that makes the LEDs turn on and off. The most common manufacturer of this IC is Macro Block.

The manufacturer of the sending card/pixel map processor and the receiving card must be the same. For example, you cannot place a Brompton processor on a wall that is using Nova Star receiving cards. When tiles are ordered from the factory, the desired manufacturer's receiving card is installed in each of the hubs. To change from one brand of pixel map processor to another would require refitting all the receiving cards in all the tiles that the company owns.

Manufacturers of sending units in order of prevalence in the industry are Nova Star, Brompton, and the latest edition Megapixel VR. As you may have guessed, the popularity of Nova Star is based on price. Hence, the large number of products that use it's sending and receiving cards.

Brompton is based out of the United Kingdom and although much more expensive than Nova Star, it has established itself because of higher quality images, ease of use, and color calibration ability. The newest player in this field, Megapixel VR, is based in the United States and is quickly becoming the pixel map processor of choice with the explosion of xR production stages and their GhostFrame technology.

The method of communication between the sending unit and the receiving cards in the tiles is Ethernet cable. However, the protocol being carried down the cable is not standard network TCP/IP or UDP communication and is proprietary to the manufacturer. This means that regular network switches and hubs cannot be used to distribute or split LED data runs. However, communication can be extended using fiber transmitters and receivers if the distance between the sending unit and the LED wall is more than 330 feet. Again, the transmitters and receivers must be compatible with the manufacturer of the sending unit.

Once the signal reaches the wall, it enters via a tile in one of the four corners and the signal is daisy-chained through the remaining tiles. Depending on the pixel count of the wall, several data lines from the sending unit may connect to the wall at different places at the point where the pixel limit for a single data line has been reached.

Power for the LED wall is handled as a separate feed, usually from a power distribution system. It does not have to enter the wall at the same point as the data, but it's a good practice to do so. Walls that are flown by truss will have the power and signal come from the top row, and walls that are ground supported will have power and signal enter from the bottom row. Again, the power loops from tile to tile and there is a limit to the number of tiles that can be powered from one circuit.

Building the LED Wall

As there are so many different connecting mechanisms and hardware between manufacturers, this section will only offer general advice based on some common practices. The first is to be aware that building the first row of any LED display will take the longest. In taking the time to ensure that your wall is level and plum will expedite the building of the successive rows to follow. For example, if the latching point between two tiles seems to take extreme force to close, then you may have some alignment error that has progressively gotten worse.

When rigging LED walls from truss, be sure to use only the manufacturer designed headers or if custom designed headers are used, that they are rated for the weight that will be suspended from them. The use of rated turnbuckles at the connection points for the headers will allow for some adjustment vertically and help with alignment. Be sure to safety off your wall with steel cables from the headers to the truss or to another rigging point above.

For ground supported LED walls, be sure to pay close attention to the manufacturer's ballast or counterweight specifications. Most ground support systems consist of a vertical ladder to which the tiles are attached and then there is a floor piece that extends backwards away from the wall sometimes nicknamed the "kicker." It is on this kicker that the ballast weight should be applied. Ballast can take many forms, the most typical is sandbags. Others include barrels filled with water or sand, or metal weights. As an added measure of safety, a ground supported wall may also be secured via overhead rigging points. Never forgo safety in order to get a LED wall closer to the back wall and always check with the company or facility providing the staging to make sure it is rated for the weight of the fully assembled LED wall.

Sourcing the LED Wall

Now that you've got your LED wall built and wired up, it's time to send it a picture. You think it would be as easy as plugging in a HDMI cable to your sending unit and having a picture magically pop up on the wall. If that were only true.

Before you assign a source to the wall, you need to map the tiles either on the front panel or through software for the pixel map processor. After all, the processor has no idea where the tiles are in relation to one another, or which one is the starting position for your data run. Once this is completed, then you can assign the correct input to that wall and get a picture, but not quite.

Depending on the resolution of your LED wall, it will only display a portion of your content, or your content with a huge area of black pixels. This is because most pixel map processors do not have scalers. Remember that a scaler is an electronic component that can resize and remapped pixels from one resolution to another. The processor maps pixel for pixel your source to the wall starting in the upper left corner at 0,0. An HD source with the resolution of 1920 × 1080 going to a wall that has a pixel resolution of 1200 × 800, will show the upper left corner of the screen with the bottom and right cut off. That same HD source going into a wall with a pixel resolution of 2500 × 1600 will show the full HD image starting in the upper left corner with a blank space to the right and bottom of unused pixels.

It is for this reason that many production companies pair their pixel map processors with some sort of external multi-format scaling device such as a Barco ImagePro or Analog Way Pulse. What if the resolution of your wall is far beyond any standard single screen resolution? In this case, you will probably have multiple pixel map processors controlling the LED wall and will need a multi-screen management system, such as Barco's E2 or Analog Way's Aquilon, to handle multi-port inputs from video servers and output discrete signals to each pixel map processor. You'll also have the ability to control the placement of other content such as PowerPoint and I-mag windows and the system will keep the outputs synchronized so there is no tearing or artifacts at the seams between processors.

Weight, Dims, Power, and Pixels

When involved in any type of LED project, these are the things that you should be aware of: weight, dimensions, power consumption, and pixel count. This information is readily available on a manufacturer's website under the technical specifications for a product. There are also many apps they have databases for multiple brands of tiles. One of these is Fido LED (www.wookiesoft.com), which is probably the most widely used of these types of apps.

Information on total weight of your LED wall is important if you're going to be hanging your wall from rigging points. The total weight should not just include all the tiles, but the headers (the hardware that tiles attach to that hangs from the truss), the weight of all cabling, and any processors or accessories that may be hanging on the back of the wall. An Absen A3 Pro tile wall with 800 tiles weighs in at 8,000 kilograms (17,637 pounds). That's over eight and a half tons!

Another thing you're going to have to get used to when dealing with LED is using the metric system. All specifications are listed in millimeters and kilograms. When companies present quotes for the purchase of product, the prices are per square meter. Let's look at the specifications for an LED tile prevalent in the rental staging market.

ROE Black Onyx 2.84

The name of the product gives us the pixel pitch, 2.84 mm. The tile size, at 500 mm × 500 mm, is standard for this range of pixel pitch products. So, if asked for dimensions and a pixel map for a wall 30 tiles wide by 10 tiles tall, our answers would be:

Table 9.1 Specifications for ROE Black Onyx tiles.

Tile Height	Tile Width	Hor. Pixels	Vert. Pixels	Tile Weight
500 mm	500 mm	176	176	9.4 kg

30 × 500 = 15000 mm = 15 m = 49.2 feet wide

10 × 500 = 5000 mm = 5 m = 16.4 feet tall

30 × 176 = 5280 pixels wide

10 × 176 = 1760 pixels tall

The next question is how many pixel map processors would it take to drive this wall? An HD processor can handle wall resolutions up to 1920 × 1200. Given the walls pixel count, it would take at least three processors to cover the horizontal resolution (1920 × 3 = 5760). Likewise, it would take two processors to cover the vertical pixel count (1200 × 2 = 2400). Three multiplied by two equals six HD pixel map processors to drive this wall.

UHD 4K processors can handle wall resolutions up to 3840 × 2160. It would take two processors to cover the horizontal (3840 × 2 = 5680) and one processor to handle the vertical. Two multiplied by one is two 4K processors.

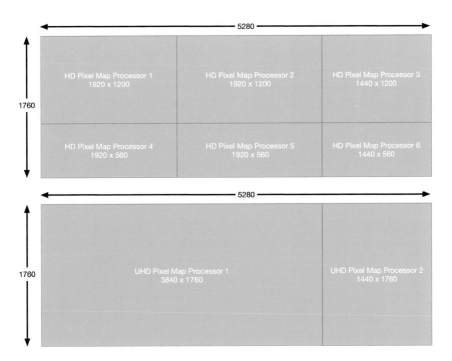

Figure 9.1 Comparison diagram of using HD versus 4K pixel map processors.

Of the two solutions, which one makes more sense? Which one is more likely to be implemented? The unfortunate answer is that six HD processors is likely to be the solution applied to this wall. HD pixel map processors are very inexpensive, and companies will have a higher inventory of those models versus the 4K model. This now narrows your choice for screen management systems or media servers since devices with six discrete HD synchronous outputs is a very narrow field of choices. However, image processing devices that can output two synchronous 4K signals would not only simplify signal feeds and reduce points of failure, it would also simplify content creation and delivery.

Currently, LED tiles are available with a pixel pitch less than two millimeters. The same sized wall that years ago had a 1920 × 1080 resolution, would have three to four times as many pixels in the same space today. LED walls of a decent size are quickly reaching, if not exceeding, 4K resolutions and savvy production companies are foregoing HD processors due to the complications they introduce in these higher pixel count walls.

Another parameter that is important with regards to planning wall configurations, is the amount of data runs required from the sending unit to the receiving units of the LED wall. We mentioned earlier that the data feed can enter in one tile and then be daisy-chained to successive tiles. Most pixel map processors have a limit of 575,000 pixels per port. Drawing on our example tile from earlier which had a pixel resolution of 176 × 176, we multiply the horizontal resolution times the vertical resolution to get 30,976 pixels per tile. Simply dividing 575,000 by 30,976, the result shows that about 18 (number rounded down) tiles can be connected to one port.

The power specification for that same tile is a max amp draw of 1.27 amps at 110 volts power and an average amp draw of .56 amps. The difference between maximum and average is that maximum is measured while the tile displaying full white. If we have a 20-amp circuit, we can loop the power through about 15 tiles (20 divided by 1.27). For ease in troubleshooting and cabling, the decision may be to split the difference and have both power and signal loop no more than 16 tiles.

Your RCFG File and You

The receiving card configuration file contains the parameters and configuration for a LED tile and its modules. The RCFG file is created at the factory and loaded into the receiving cards when the tiles are assembled. There have been cases where the firmware in the tiles had been upgraded during production and shipments of tiles were sent out with both old and newer RCFG files.

There are two ways to recover corrupted or missing RCFG files for LED tiles. If you have a good tile, you can export the RCFG file from that tile to your computer and then download it to the problem tile or tiles. Also, most manufacturers have a download site for the factory default RCFG file for your product.

RCFG files are unique to the manufacturer and model of the tile. A file from the same manufacturer but for a different product will not work. As you work with more LED brands, it's a good idea to keep a library of known good RCFG files. This way you won't be scrambling for an internet connection in the middle of the night trying to find a file.

LED Viewing Distance

LED panels are specified by pixel pitch, which is the distance from the center of one LED module to the next expressed in millimeters. A pixel pitch of 10 mm to 20 mm is common for outdoor displays, while pitches of 6 mm or less are found on indoor displays. The viewing distance of the audience can be used as a determining factor in what pixel pitch to specify.

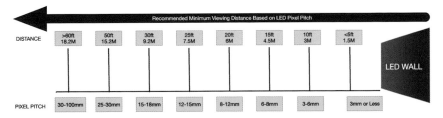

Figure 9.2 Comparison of viewing distance to LED pixel pitch.

As you can see from the above diagram, the pixel pitch required increases in correlation to the distance of the viewer. The minimum viewing distance is the point where the viewers eyes cannot see the space in between the LED pixels and the image appears continuous. So, if the minimum viewing distance of current 2.9 mm tiles is about five feet, why is it used when the audience is often much further away?

The answer may be more psychological than physiological. The concept of smaller pixel pitches plays into the emotion of wanting to have the latest and greatest technology. Sales and marketing teams at LED manufacturers would be more than happy to tell you that smaller pixel pitches equate to better image quality and in the sense they wouldn't be wrong. A wall with a 2.9 mm pixel pitch would look far better than the same size wall with a 6mm pixel pitch.

Another reason may be the expectation of the end users. With an LED wall being the main stage set piece as well as the main presentation display, the reasoning for the small pitch becomes the ability for the presenter to read the material from the screen while on stage. They are usually three to five feet away and the concept in their mind may be that "If I can read the content clearly from where I am, the audience certainly should be able to read it." This idea could also be extrapolated to considering the viewers in the first few rows of the audience. Even though they may be 20 to 30 feet away from the display, this is where the VIPs and executives whose companies are paying for these shows are sitting.

It returns to the basic economic principle of supply and demand. The manufacturers of LED tiles are producing indoor product with a sub-3 mm pitch because that's what's being requested of them, and they are capable of it. Their dedicated outdoor products will probably stop at around the 6mm or 8mm pitch because smaller pitches for outdoor viewing are pointless.

The Holy Grail of LED products is a display that has the clarity of projection, the brightness of LED, and the ease and setup of a fast-fold projection screen.

Moiré, Flicker, and Friends

LED is not without its issues. When shooting or filming LED on camera, there are a number of things that can detract or be unpleasing to the viewer. Based on the quality of the processing, these can be quickly dealt with or remain unsurmountable obstacles.

Unprocessed, LEDs have different frequency or refresh rates depending on the color and brightness they're trying to produce. When used as the backdrop on set, the amount of flicker produced by the LED wall may vary from color to color and be very distracting. It will not be seen by the naked eye, but the difference between the refresh rate of the camera sensor and the frequency of the LED causes this phenomenon. The higher quality LED manufacturers who sell into broadcast and film markets, have a means to deal with this. By setting a constant refresh rate, such as 3840 Hz, that remains the same regardless of what is being displayed on the wall, they can greatly reduce flicker.

When using any lighting or creative LED products that do not have this constant refresh, the camera engineer and the lighting director will have to go through the scenes and colors to determine which colors have no flicker and which colors cannot be used. Lighting that is driven by LED sources will have the same issue and again, and some may have modes for use on camera that provide a constant refresh rate.

The other aberration that may occur when shooting LED walls is called moiré. Moiré patterns appear when the pixels of the LED wall do not line up exactly with the pixels of the camera's imager. Appearing as symmetrical curves of shadows, there are a few methods that can be tried to reduce this effect. The first is to try changing the distance between the LED wall and the subject being filmed. Increasing the distance may blur or reduce the effect by changing the focal distance between the wall and the subject. Second, and quite commonly used in broadcast, is to place a scrim-like material just in front of any LED panels that will be on camera. This defocuses the LED pixels to the camera and eliminates this effect.

Finally, reducing the brightness of the LED wall allows the camera operator to use a larger iris setting which in turn creates a narrower depth of focus, blurring the background. This last solution may only be viable in studio situations where you are shooting content and not necessarily in a live event in front of an audience.

And Now a Word About Virtual XR Production

With the COVID-19 pandemic of 2020 shutting down most live events, conventions, and television and film production, not only did virtual XR (extended reality) production find new use cases, but exploded as a production methodology.

A typical virtual XR set involves a curved LED wall usually 10 to 12 feet tall and 40 to 50 feet round. There is also an LED floor and ceiling to which content from media servers or game engines are sent. This setup is referred to as a "volume." Capturing images in this volume, is a cinema camera mounted on a jib or robotic crane. It's movement and position in XYZ space is calculated by a tracking system that follows dots on the floor and ceiling or uses other cameras to track the main camera.

The result is an environment in which actors can interact with the locations or interior spaces that their scene takes place in. As the camera follows the actor's actions, content on the wall behind them, an area called the "frustum," updates in real time to match the size and focus of the background as if they were shooting in the real world. The LED displays provide lighting for the scene and can include real time reflections such as clouds passing through the sky on the windshield of a car.

I share with you this information about virtual production not because it relates to live events per se, but because someone with skills in LED technology, media servers, and cameras could find a career path in a virtual production environment that utilizes a lot of the same skill sets.

In the live event space virtual production is relegated to taping segments that will later be used in a live event or as a backdrop to a traditional multi-camera shoot.

10

NETWORKING AND THE INTERWEBS

Congratulations! You just purchased a new computer and when you get home, you turn it on and begin the configuration process. The laptop automatically finds a list of available Wi-Fi networks, including your personal network at home. It asks you for the Wi-Fi password, which you enter, and the setup program continues to download the latest files for installation from the Microsoft or Apple servers.

Have you ever wondered why all that was needed to get to the internet was a password? Likewise, when you go to Starbucks or stay at hotels, all that is required to go online is a password. How does the Wi-Fi network distinguish you from all the other computers? Is the same information used at home used at Starbucks to get you online?

The truth is, in everyday life the need for people to know the details of their internet connections is not only unnecessary, but for the most part useless information. However, for show control networks, the ability to establish and configure IP addresses manually is a crucial skill.

DOI: 10.4324/9781003247036-10

How networks transmit data and interact with applications is laid out in the open systems interconnection (OSI) model, or what is also referred to as full stack. In this model, there are seven layers to the transfer of data over networks and between hosts. For our purposes we will not be going in depth into this seven-layer model. You're more than welcome to dive deeper into the subject matter in an online networking course or computer science class.

Connecting to a network begins with an NIC, network interface controller. This can be an integrated component on a motherboard or added to a system through a card slot. Every NIC is assigned a MAC address. This is not an address that is controlled by Apple, but a unique serial number for the network hardware that is the same no matter where you are or what network you connect to. It is like a fingerprint in that there are no two identical MAC addresses. The hardware MAC address is included in the header of the data packets that travel across networks or the internet.

Internet protocol, or IP addresses, are numbers assigned to devices on a network and can change from network to network. These devices, referred to as "hosts" or "nodes," are usually part of what is called a LAN, or local area network. A WAN is a wide area network that is used to describe a large grouping of integrated networks, such as in a corporate headquarters or on a school campus. A WAN may be made up of several smaller VLANs, which is short for virtual local area networks. But maybe we're getting ahead of ourselves.

An IP address is a 32-bit number made up of four octets and looks something like this: 192.168.1.102. The spaces before and after the dots are made up of 8-bit numbers, thus octet. An IP address defines two pieces of information, the specific network and the host. In our example, the first three octets, 192.168.1, are the network, and 102 is the specific host on that network.

Working in Binary

So how do we create 8-bit numbers? Each of the bits has a numerical value.

Let's take our example IP address, 192.168.1.102. The 192 in the first octet is expressed in binary as 00000011. Only the last two bits are turned "on," indicated by a 1, all other bits are "off," indicated by 0. Bit 7 (64)

Table 10.1 Numerical value of each bit in an 8-bit system.

Bit 1	Bit 2	Bit 3	Bit 4	Bit 5	Bit 6	Bit 7	Bit 8
1	2	4	8	16	32	64	128

plus Bit 8 (128) equals the number 192. Each bit can only be used once, and all numbers can be represented up to the limit of 255.

Proceeding through the remaining octets:

$$168 = 00010101 \quad (8 + 32 + 128).$$
$$1 = 10000000 \quad (1).$$
$$102 = 01100110 \quad (2 + 4 + 32 + 64).$$

Therefore, the IP address expressed in binary would be:

00000011.00010101.10000000.01100110.

Congratulations, you can now count in binary and speak in the native language of computers. Can the number 343 be used in an octet? No, because it cannot be represented by 8-bits. How is zero or nil represented in binary?

The organization that handles the allocation of IP addresses is the IANA, the Internet Assigned Numbers Authority. There are also regional internet registries, RIRs, that receive blocks of addresses from the IANA. Two protocols exist for IP addresses, IPv4 and IPv6. The IP address used as our example and most IP addresses you may have seen are IPv4. Believe it or not, there are 200 billion combinations of IP addresses available under IPv4, and they all have been allocated.

Whilst IPv4 had a 32-bit size for addresses, IPv6 has 128-bit addresses and uses hexadecimal numbering and looks something like this: 2001:db8 :3333:4444:5555:6666:7777:8888. At four times the available IP addresses, we should be covered for a while.

Keep in mind that IP addresses are only assigned to those devices connected directly to the internet, such as routers, IP phone systems, and web servers. Networks built on show site will use what are called non-routable or private IP addresses. Generally, these address start with 10., 192.168., or 172.

These non-routable IPv4 addresses can only be seen within the local network. They are usually distributed through a network router by a

DHCP server. Dynamic host configuration protocol automatically assigns IP addresses to devices on the network that request one. Most internet capable devices, such as laptops, smart phones, and tablets, have DHCP addressing turned on by default. This is what makes it so easy to connect when you were at home or at your local Starbucks. Your cable router or the network router at a location will assign you an address that doesn't conflict with any of the existing devices on the Wi-Fi or wired network.

For show control, it is necessary to have fixed IP address assignments. That way if there is an issue or problem, you know exactly which device is causing it and physically where it's located. More often than not, these networks will be set up with the 192.168.xxx.xxx scheme. You can have up to 255 devices on a network. If you need more, you can increase the number of hosts through subnetting.

When reconfiguring your network card from DHCP, automatic addressing to manual, you will be asked to enter the IP address and subnet information. Subnets have four octets just like the IPv4 address. A typical subnet address looks like this: 255.255.255.0. The 255s indicate the network portion of the address and the zero indicates the host value. Networks have an easier time transferring information between hosts if they are told that they only have to resolve the last octet of the IP address when looking for another host on the same network. The zero in the subnet limits the search to 254 devices. To increase the range of hosts in a network of private IP addresses, the third octet of the subnet can be modified to allow more hosts. A subnet of 255.255.252.0, would allow 1022 IP addresses on the same network. The third and fourth octets could be used to identify a host. An IP address and its subnet are expressed as: 192.168.0.102/24. The 24 after the slash indicates that 24 bits, or three octets, of the subnet address are 255: 255.255.255.0.

A subnet of /16 would be: 255.255.0.0 and a subnet of /22 would be 255.255.252.0.

A final field you may see when entering manual IP addresses is a place to enter a router address. Since for practical reasons your show control system should not be accessing the internet, there is no need to enter a router address in this field. If the computer is persistent about wanting an address, you can simply type in 192.168.0.1 or whatever matches your network scheme.

So how do I change my network IP address from DHCP to a manual fixed address? Since there is the possibility that someone reading this maybe doing so during a major release of Windows or Mac OS, you will get very generic directions. On Mac OS, click on the Wi-Fi icon in the top right of your finder toolbar. Scroll to network preferences and click. You could then select the connection type, Ethernet or Wi-Fi, and then change the IPv4 configuration from using DHCP to manual and fill in the parameters. For Windows users, click on the search in the bottom left of the taskbar and then enter network adapter. There should be a selection for change network adapter settings. Click on properties, and then IPv4 settings and enter the information. Don't forget to change back to automatic or DHCP setting when the show's over and you go back home or to your hotel or you won't be able to get on the internet.

So now that you're armed with the information about what an IP address is and how to set one on your laptop or computer, you can create a network topology for your show. Topology diagrams can be as simple as a block diagram that shows all the devices connected on your network. It can include tables that show what IP addresses are assigned to what pieces of equipment.

There are several types of hardware that can be used to interconnect host on a computer network. These are routers, switches, and hubs. Each of these has a different purpose for being part of a network topology. Hubs allow multiple computers to be connected together and communicate on a network, but there is no management or traffic control of the data. Data packets often have "collisions" as they go from one host to the other looking for the correct recipient. Imagine a train station where there are several trains sitting on the tracks, but no signs to indicate where the trains go. Passengers would need to go up to each train and ask what their destination was, and in that process, constantly collide with other passengers trying to get the same information. This is how data flows through a hub.

Now imagine the same train station with an information board that shows the track number and destination for each of the trains. This is how a network switch operates. It has a small domain server that remembers which hosts are on which ports and can direct the request from a host to the proper port, avoiding collisions. Managed switches go a step farther by allowing the user to set up small VLANs where certain ports

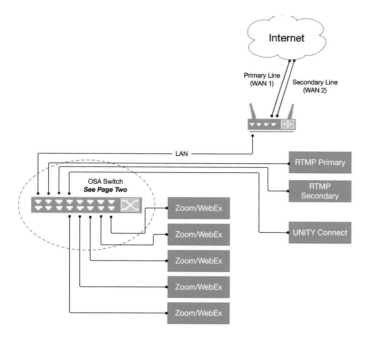

Switch Details

Port 1	Port 2	Port 3	Port 4	
From Venue	RTMP Primary	RTMP Secondary	UNITY Connect	
4 Public IP's +DHCP 125Mbs Bandwidth	TCP Port 1935 DHCP 10 Mbps	TCP Port 1935 DHCP 10 Mbps	TCP Port 20102 Static Public IP 10 Mbps/Server	
Port 5	Port 6	Port 7	Port 8	Port 9
Zoom / WebEx	Zoom / WebEx	Zoom / WebEx	Zoom / WebEx	Zoom / WebEx
DHCP 3 Mbps/Meeting	DHCP 3 Mbps/Meeting	DHCP 3 Mbps/Meeting	DHCP 3 Mbps/Meeting	DHCP 3 Mbps/Meeting

Figure 10.1 Sample of a network topology diagram.

are restricted to communication between certain hosts or groups of hosts. Several VLANs can exist on the same managed switch.

A router's job is to communicate with other routers on the internet and are the gatekeepers for access to hosts on their local network. Routers typically also have the job of being DHCP servers and assigning IP addresses to hosts on their network. Going back to our train station analogy, the information boards only know the track numbers of the trains that are local. A router adds trains that that connect to other cities across the country.

Configuring a network to control several projectors for the purpose of alignment is a typical show site scenario. Often a wireless access point will be added so that the technician can walk around with a laptop or tablet wirelessly. Although convenient during setup and rehearsals, the wireless access point could become a liability during the actual run of show. It is often recommended to disable the wireless and go with a hardwired Ethernet connection during show days. This is due to the fact that every attendee will probably have a Wi-Fi enabled smart device or two on their person or in their bags. Those devices will constantly be pinging for internet access and the traffic could slow down your ability to send commands to projectors or block them completely.

Firewalls may need to be configured or disabled in order to gain access to communication ports needed to control devices. There should be a very low security risk in doing so if the show control network has no access to outside wide area networks or the internet.

Video Over IP

Over the years there have been many protocols developed to transmit video or combinations of video, audio, and control over Ethernet cable infrastructure. The reasonings have been that Ethernet cable is inexpensive and as part of most buildouts in the corporate and educational sector, Ethernet cable is run everywhere. Let's explore some of these technologies in a somewhat chronological order of appearance.

HDBaseT

This is a point-to-point transmission system using Ethernet cable as the conduit. Dedicated interfaces at either end convert audio, video, and control signals into data packets that are transmitted up to 330 feet or 100 meters between devices. The system can except HD or UHD 4K video sources and many projectors and industrial monitors have HDBaseT inputs directly on them. The specification also allows for TCP/IP networking protocols to be part of the signal. With one Ethernet connection into the projector, you can send a high-quality video signal and be able to control the projector through direct communication or access to built-in web browser control.

Again, it is point-to-point so you cannot use network interfaces such as switches or hubs to boost or distribute the signal to multiple receiving devices. There are dedicated HDBaseT distribution switches should a single source need to be sent to multiple destinations. HDBaseT is used mostly in the fixed installation and integration markets.

Video Baluns

Another point-to-point device, baluns are low-cost ways to extend the distance of certain signals such as SDI or HDMI using Ethernet cable. Mostly relegated to consumer or non-professional uses, baluns can be an inexpensive piece of hardware to keep in your toolbox for an emergency situation.

Audinate Dante

Although not a video transmission method, it is important to mention the Dante audio over IP system and how it changed audio signal transmission. Developed by Audinate in Australia, Dante converted analog or digital AES audio signals into network data packets that could be broadcast over a computer network. Multiple channels of audio could be sent or received over a single Ethernet cable and off-the-shelf network hardware. Using Dante network management software, users could detect all of the Dante compatible devices on their network, and assign which signals should go where with simple matrix routing. It simplified and reduced the size of audio snakes and the number of individual patches that had to be made in a sound system. Generally, audio does not require a large amount of bandwidth for data. Conversely, video requires huge amounts of bandwidth, and the hopes of a Dante-like system for video were desired.

SDVoE—Software Defined Video Over Ethernet

The SDVoE Alliance was founded by a group of AV manufacturers, system designers, integrators, and technology managers in order to replace point-to-point systems with matrix switch versatility. Unlike HDBaseT, SDVoE uses off-the-shelf networking hardware to route and distribute signals carrying AV information. It eliminates to need for bulky and expensive matrix switches with are format specific and usually limited to a handful of manufacturers.

Like HDBaseT, hardware is required at the send and receive endpoints in order to convert HDMI video, analog audio and control (IR, TCP/IP) into data with an IP address, that can be put on the network. Assignment of IP addresses is handled through a system management program that can detect all compatible devices on the network. Minimum bandwidth required is 10 Gps and the receiving devices just need to be assigned the IP address of the video they want to display. Since the data is multicast, multiple receivers can access the same IP address simultaneously.

Another feature of SDVoE is how it encodes video for low latency, which is ideal in live situations. Using pixel pipeline compression, each line of video is encoded and transmitted versus encoding entire frames like the H.264 codec. This reduces the amount of compression needed to 1.3 to 1, versus the 200 to 1 with H.264. Also, the delivery time to the end device is just a few lines, or just a few hundredths of a millisecond.

NDI

Network device interface (NDI) not only provides a means for video over IP transmission, but is an ecosystem of products that manage video over IP as a source and output when producing programming for live events or the web. Unlike SDVoE, which is primarily a means of distributing content over Ethernet using hardware endpoints, NDI adds the convenience of having the protocol built-in to products or available as a software plug-in. It also provides endpoint solutions for devices that need conversion to NDI and IP addressing is handle through management software.

Dr. Andrew Cross, who was developing IP video technology at NewTek, introduced NDI in 2015. It was primarily an IP video tool to support their

TriCaster production switchers and was built natively into the platform. In the following years, the technology would lead to smartphone apps that allowed them to become NDI camera sources, integration into PTZ cameras for not just video transmission but movement control, and integration into vMix and OBS, the most popular software switchers for web-based production.

NDI's royalty-free technology was being added to converters and programs that could capture computer desktops and make them NDI streams. What made the idea of NDI popular as a way of sourcing video was the scalability. With HDMI and SDI, once you are out of physical inputs you need a bigger switcher. NDI allowed as many compressed streams as could fit into the bandwidth of a given Ethernet connection, and with network multicasting there was no need for video routers or distribution amplifiers.

NewTek would be acquired by Vizrt Group in 2019. Based in Norway, Vizrt had a long history in the broadcast industry of interactive graphics and augmented reality. In 2021, Vizrt announced their cloud NDI production service. Picture a cloud-based TriCaster where content could be sourced from multiple locations in real-time and switched from a central location or remotely.

SMPTE ST-2110

What Dante is to audio, ST-2110 is to video. The approach of Dante was to replace the infrastructure wiring of an audio system with an Ethernet network-based topography. Likewise, ST-2110 is the replacement of SDI transport in a video production system with Ethernet.

Video in ST-2110 is uncompressed, format independent, and is multicast across the network using Class D private IP addresses 224.0.0.0–239.255.255.255. ST-2110 transports video, audio, and data in separate packet streams called "essences." Like signals in a traditional SDI system, they require a reference sync signal in order to maintain synchronization. With networks, the packets are timestamped and PTP (precision time protocol) is responsible for this. A central master clock device, such as the Evertz 5700MSC-IP or Tektronix SPG8000A, can provide network master clock as well as synchronous analog black burst outputs for legacy equipment.

The standard was developed to be part of an equipment's specification and built into the product without the need for external converters. There are a number of professional cameras that have ST-2110 options for output modules and most major brands of broadcast production switchers have models with ST-2110 operability.

The heart of the system is the network switch. Given the data rates of HD and 4K video, the network bandwidth required starts at 10 Gbps and should probably just be 100 Gbps to allow for growth and the flexibility to handle 4K events. A 3G SDI signal is still 3 Gbps of data in ST-2110, so your typical gigabit Ethernet switch will not work to transport even one stream.

One of the advantages of using existing network technology, is that video systems can grow and expand by adding and linking hardware just like data centers do with servers. However, you also have their price tags. A 16-port 100 Gb Ethernet switch can cost anywhere between $5,000–9,000. Compared to a simple 16-port Gigabit switch from your local electronics store for less than $100. The cost of a 40 × 40 12G-SDI matrix router is about $6,000.

One facet of the ST-2110 system that is unique is built-in redundancy under the ST-2022-7 part of the specification. What this means is that a ST-2110 source that has main and redundant streams can take two different network paths to the switcher. Should one of the streams encounter an issue, the switcher can automatically go to the other stream. This may be more useful for signals that originate remotely, versus within the studio or venue.

When making connections at these higher data rates, you will be using a device called SFP, or small form-factor pluggable. More specifically, SFP+ since these are the modules and ports rated for 25 Gbps and above. An SFP+ module can be outfitted with copper or fiber-optic cable, or ones used for ST-2110 connectivity may have HD-BNCs right on the front of the module for converting SDI to networking protocol. A pair of 25 Gbps SFP+s might be all that is needed to ingest all your ST-2110 sources into a compatible production switcher.

How many streams can be carried on Ethernet in ST-2100? This chart shows the maximum number of streams per given network bandwidth not including redundancy.

Table 10.2 Maximum number of ST 2110 streams per port of network band width without ST-2022–7 redundancy.

Standard @ 59.94	Rate Mb/s	1 GbE	10 GbE	25 GbE	40 GbE	100 GbE
HD 720p- SDI	970					
Uncompressed ST-2110	863	1	11	28	46	115
HD 1080p- SDI	2,970					
Uncompressed ST-2110	2,635	0	3	8	13	34
4K UHD 2160p-SDI	12,000					
Uncompressed ST-2110	10,466	0	0	2	3	8

Be aware that ST-2110 supports formats and data rates from standard definition SDI through UHD 4K. These numbers will vary with a mix and match of formats on the same port.

SRT—Safe Reliable Transport

Developed by the company Haivision, SRT (safe reliable transport) is an open-source, point-to-point video transmission system for use on public broadband internet. It allows for low-latency, high-quality images using the UDP data transfer protocol.

In data transfer, there are two protocols: UDP (user datagram protocol) and TCP (transmission control protocol). UDP is faster but is primarily for pushing data without acknowledgments or error checking of the data. TCP is slower, but makes sure connections are established and checks the data for missing information. SRT uses the faster UDP protocol, but puts a layer of packet loss recovery and encryption on top of it. Thus, more reliable transport.

The SRT transmitter and receiver at each end must have a fixed public IP address. This is usually assigned by the internet service provider at each location. Each interface can process four discrete channels of 1080p video or one quad 4K. AES audio can be embedded/de-embedding on any of the four channels.

A use case for SRT is a studio with multiple cameras being sent back to a control room in another city. There the program is switched and sent via SRT to a network master control where it is broadcast via satellite worldwide.

RTMP

Real-time messaging protocol was originally developed as a TCP based technology for Adobe's Flash Player. Today, with the prevalence of streaming to Facebook, YouTube, and wherever else, its role is to deliver content from encoder to an online video host.

It is low-latency and supports H.264 as the preferred encoding codec. There is not too much configuration required as long as you have a quality streaming service and software or hardware encoder such as OBS, vMix, Wirecast, or the Epiphan Pearl to name a few. With the end of Adobe Flash and its support, it will be interesting to see where this protocol goes.

HLS

HTTP live streaming based on HTML5, was developed by Apple to stream video and audio content using the same protocol as most popular internet browsers.

It is dynamic in that it optimizes playback based of the speed of the wired or wireless connection. Video and audio are compressed into HEVC and AC-3 audio respectively and output as a MPEG-2 transport stream. The video is then chopped into segments and sent. The segments live on a server in a playlist, and when played out the viewer sees the stream as continuous with no gaps.

Early version of HLS has a latency of 10–12 seconds from source to viewer. This was detrimental to live events such as sports, gaming, or anything with a real-time social component. Low-latency HLS has reduced this to 1–2 seconds in recent versions.

11

THE INTANGIBLES

So, you've decided to be a video technician. Good for you! You have learned the latest switching systems and taken workshops on the latest projectors and LED panels. What now? Now you make sure to take care of the intangibles.

What are those? Intangibles are the things you can't see or touch. They are the personality you bring to the show site. They are how you treat other people and the respect you show them. The things that make you someone to call on again and again, versus people wincing when your name is mentioned.

Punctuality

"Being early is on time, being on time is late."

This is a mantra you will encounter ad nauseum. Fifteen minutes before the call time is the generally accepted rule for being considered "on time."

DOI: 10.4324/9781003247036-11

Thirty minutes is better. But why show up early if you're not getting paid? This pre-call time is used to check-in with the union steward or the production manager, grab some coffee from craft services, or have that last bathroom break. At call time, work begins. Showing up just as work starts gives you little transition time and the last thing you want to do is have a crew waiting on you.

This is about respecting other people's time. If you do show up late after the call time, you better have doughnuts. Take into account not only the time to get yourself ready, but the drive time and time from the parking lot to the venue or ballroom. In Las Vegas for instance, it can be a good thirty- to forty-five-minute hike from the parking garage, through the casino to the meeting rooms. Even if you are staying on property, a large resort hotel will have your room the farthest possible distance from the ballrooms.

We are all human and life happens. If you get a flat tire or delayed because of an accident, let someone know and that you're on your way. Getting flat tires every day, however, will arouse suspicion. This does bring up an interesting point. When you get booked for a gig, try to get a contact number for the department lead, steward, account executive, or someone else on the crew in case you have to alert them of any situations.

What Do I Bring?

First of all, bring yourself, both physically and mentally. Once that is completed successfully, there are a few items that always should be in your backpack or toolkit that goes with you to a show site.

A good multi-tool like a Leatherman or Gerber. You may be thinking old school and go for the traditional Swiss Army knife, but there are probably more things on there that you are never going to use. The multi-tool gives you the four main players: sharp knife, pliers, cross-point (Phillips-head), and flat-head screwdrivers. Be sure to pack it in your checked luggage if flying to and from shows. The TSA will confiscate it since you can't take in on the plane because of the knife and pliers.

A six- to eight-inch adjustable wrench—what our friends across the pond refer to as a "spanner." This is used for adjusting projector rigging cages or tightening nuts. Be sure to have a leash and carabiner clip on it for when you're up in a lift or on a ladder. Accidentally dropping this could ruin someone's day.

A flashlight. Not so much during setup and "strike" (the nickname for taking down a show), but for the times in between when it is dark. It doesn't have to be one of those tactical 10,000 lumen LED jobs, but a good mini-Mag size flashlight you can have on a waist clip. A headlamp option is also good to have for wiring racks or loading/unloading trucks.

Black Sharpie markers. Lots of them. Use a Sharpie to write your name on your Sharpies. Hang one from your lanyard or show credentials. Have one in your pocket (remember to remove before wash day).

White board or paper tape. The companion to your black Sharpies, this tape has a non-residue adhesive and you can use it to mark the faders on an audio console, label video recorders and switchers, and most importantly, labeling the ends of every cable that you run. Nothing makes more of an impression than for an engineer to not have to guess which cables are which because they were properly labeled.

A BNC removal tool. Also referred to as a trompeter tool or "colon scraper." This tool has a handle like a screwdriver, a foot-long shaft, and a circular collar at the end. This is used for removing or installing BNC connectors on equipment buried in racks. Since the BNC ends have to be turned in order to lock, this tool makes it easier when they are packed close together and you can't get your fingers in around the connector. Be sure to also have the HD-BNC version as well since it is more commonly being used to fit more connectors on a smaller chassis.

A simple electrical outlet tester. These can be found in the electrical department of your local hardware or big box store. It is a little yellow plug with three lights on it. When plugged into an electrical outlet, all the lights are illuminated signaling that the wiring in the plug is fine. If any of the lights don't light up when plugged in, it indicates a bad or missing connection. The worst of these is a missing ground connection. If a device does not have a ground connection, there is a risk of electric shock since it is a short to ground that "trips" a breaker and kills power. Use the tester to check electrical outlets before plugging in expensive electronics.

If you're ready for the next level, having the next two tools should really impress. An Ethernet tester. This is a device that has a detachable box that goes on one end of an Ethernet cable. The other end of the cable goes into the tester and it will show you if all the pins are connected correctly. A great way to test if an Ethernet cable is good before running it long distances and a way to locate a cable from among many in a pile.

The other is a multimeter. This device measures amps, voltages, and continuity for AC and DC power sources. You would use it to measure voltages at the end of projector runs, verify correct voltages from a power drop, or just test signal passage from one end of a wire to another. When engineering camera tallies, the multimeter can indicate a successful contact closure from the switcher. Since power measurements need to be accurate and your life is at stake if the meter readings are wrong, spend the extra money to get a meter from a well-known company such as Fluke. They also have accessories for your meter, such as amp probes that you clamp onto to a power cable to measure the amp draw.

What's for Chicken?

For most eight- or ten-hour days, assume that lunch will be an hour and it will be a walk-away. In other words, you are on your own for lunch. If lunch is only going to be half an hour, the meal should be provided for you since it is unreasonable to think that you can go away, eat, and return within 30 minutes.

When catering is provided, the protein of choice is usually chicken. It is universal in its flavor and non-offensive to everyone except vegetarians and vegans. This is usually accompanied by a pasta or potato-based side. If you have special dietary considerations, be sure to pack your lunch or make the best of what's served. This is said not with ill will as in your mother telling you to stay at the table until you eat all your vegetables, but as a warning of what to expect from most hotel catering.

If you are show crew or you are going to be onsite during the run of show and participating in catered lunches, please don't be afraid to mention being vegan or allergic to glucose to the technical director, steward, or whomever is placing the lunch orders. They may be able to get you a special plated lunch or if there are several on the crew with that request, something added to the offerings. They may do nothing at all to accommodate, but at least you can ask.

The other thing about catered meals is to go easy. Don't run to be first in line and pile scoopfuls onto your plate. Moderation is important, as the food has to feed everyone. If everyone has gone through and there is extra, then by all means get seconds. Along those lines, if you show up for a strike call and there is catered food sitting there, don't touch it. More than likely, it is dinner for the operating crew that hasn't gotten up from their positions in hours. Nothing pisses people off like watching incoming crew eating their highly anticipated dinner.

What to Wear? Black, Black, or Black?

There is no doubt that after some time your wardrobe will become a singularity of fashion. A black hole. Did you know there are actually different kinds of black clothing? There are show blacks, such as black jeans or pants with a black collared shirt or polo. Work blacks, such as black cargo shorts or jeans with a dark gray or black t-shirt. Blue jeans and tan cargo shorts may be substituted on the first day of load-in, maybe also day two. By day three or rehearsal days, your shades should be getting considerably darker.

The safe bet when showing up to show site for the first day is black jeans and a black t-shirt. Always be sure to wear comfortable shoes. You are going to be walking and standing for hours.

Depending on climate and time of year, shorts may be appropriate up until rehearsal days. Never shorts that are torn or too short. Some hotels and resorts have dress codes for production personnel working on their properties that have to traverse back of house areas or hotel guest areas. The production company or technical directors should make you aware of these requirements so you can pack accordingly. It may be hard to enforce if you are also a guest staying on property.

These days, health and safety on show sites is a great concern. If you are going to be working in areas where active rigging is taking place or during a load-in in a stadium or sports arena, you may be required to wear a hard hat and high-vis (neon) vest. Have those with you on the first day just in case. Also, compete an OSHA 10-hour safety training. These can be done online and the cost varies. You usually can complete the material in two days.

Now since this book is being released in a post-COVID-19 pandemic world, the following may be relevant, or not, based on prevailing CDC guidelines and positive test rates. If you are able to receive a vaccine, get one. The trend at the time of this writing is that those technicians that can immediately show proof of vaccination are going to be booked before those without. It's a hard truth. There are countless stories of people being booked and when the end client or venue requires vaccinations, production companies scramble to find replacements.

Here is an example of current large meeting protocols. Production personnel must have a negative COVID test result with 48 hours of arriving onsite, vaxed or not. Once onsite, they are required to take a rapid test prior to reporting to work and every other day throughout the run of the event. Any positive test will require immediate quarantine in their hotel room until an evaluation can be made by a doctor or other medical personnel. If they are local, they are immediately set home.

How does a production company staff for this? It requires having techs on stand-by in case one of their crew tests positive and for at least as long until a negative result can be returned. Does the original technician get paid for the days they miss if they end up testing negative later?

One event even went so far as to hire an "A" crew and a "B" crew that mirrored all the positions. The A crew loaded in and ran the show unless someone tested positive in daily screenings. If someone was positive, the A Crew went to their rooms and B crew, who had been quarantined in their rooms until this point, came out and ran the rest of the show. That's extreme, and one has to wonder how a crew can take over a show without participating in rehearsals or what happens if the B crew has a positive result?

The fact we are acknowledging is that there remains a risk, and how do we mitigate it and still continue to be able to meet face-to-face?

A Typical Production Schedule

Here is what a typical load-in schedule looks like:

8:00 a.m. Crew all time.
8:00 a.m. to 9:00 a.m. Unload trucks and push gear to place.

9:00 a.m. to 10:00 a.m. Build screens. Layout tables and equipment racks.

10:00 a.m. to 10:15 a.m. Break.

10:15 a.m. to 1:00 p.m. Hang projectors on truss. Run power and signal cabling for projectors. Power system racks. Establish camera positions.

1:00 p.m. to 2:00 p.m. Walk-away lunch.

2:00 p.m. to 3:30 p.m. Continue wiring racks. Run cables to stage for downstage monitors.

3:30 p.m. to 3:45 p.m. Break.

3:45 p.m. to 6:30 p.m. Test signals and feeds. Troubleshooting.

7:00 p.m. End of day one.

Day two may mean more setup for larger events, but typically you have another half a day to finish items from the previous day and get ready for rehearsals later in the afternoon.

Getting Paid

Probably the most important part of any gig is making sure you are rewarded for your efforts. If you were hired by a labor broker or a union, make sure the onsite lead or job steward has your hours recorded. Keep track daily and make sure the times you have written down match what is going to be submitted.

If you decide to freelance, compare your hours with the show's technical director and send your invoice as soon as possible after the show. Most good companies will pay within two weeks. Others will take up to 30 days or more, so make sure have a financial plan if someone pays late, or worst case, not at all.

Your rate for the show should have been agreed upon when you booked the show and be sure to get this in writing (paper, email, text). This will help if there are issues when you invoice. Remember, your invoice is a bill, not a suggested donation.

At the End of the Day

Being a live event video technician should be a job you love as much as the people you work with. You will make friends from various places and encounter various point of view and ideas. In the hotel bar at night, you will have conversations and share stories and despite any differences find time to laugh, share the pain of being away from families, and create bonds that last a lifetime.

You may also be lucky enough to have a position which keeps you in town every night at a local venue or production, able to attend school plays, and have date nights on a regular basis.

Entertainment and live events are full of diverse, hardworking people. Since audio, video, and lighting technicians can be found on every continent, and we all use the same equipment, there is a universal bond. At the end of the day, age, gender, sexual orientation, race, religion, and political affiliation are all put aside for the sake of the show. If we do our jobs correctly, the audience won't even notice we are there. Just a seamless part of the experience.

INDEX

Note: Page locators in **bold** indicate a table. Page locators in *italics* indicate a figure.

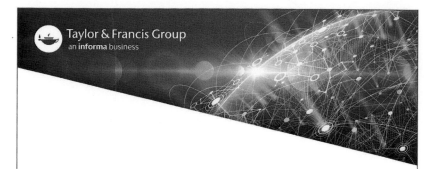